COLORADO

PLACERS

AND

PLACERING

The picture on the front of the book is of a gold mine and tailings in the mountains near Breckenridge, Colorado.

reprinted 2001 by

Carson Enterprises, Inc

P.O. Box 716

Dona Ana, New Mexico 88032-0716

1-505-524-0268

ISBN 0 941620-61-1

CONTENTS

FROM CARSON ENTERPRISES--

(Reference Library Catalog and dealer discounts are available.)

Other books along this same subject line are:

Gold! The A B C's of Panning by Rocky LeGaye..$8.95

Extraction of Free Gold by Al McGowen..$6.95.

Principal Gold Producing Districts for:

Colorado........................5.95
California........................5.95
Idaho..............................4.95
Oregon...........................4.95
Alaska............................5.95
Arizona...........................5.95
Utah...............................4.95
Nevada...........................5.95
Montana..........................5.95
Washington......................4.95
New Mexico......................5.95
Mich., S. Dak., Wyoming..4.95
Ala., Ga., Va., Tn.,
and Penna.................4.95

INFORMATION ON TERMS USED IN THIS BOOK

COMPOSITION

Placer gold changes in composition with time and as it is transported. All placer gold is alloyed to some degree with silver and sometimes other metals. The placer gold produced in Colorado has ranged from 700 to nearly 900 fine and some from Douglas County has been exceptional 990 fine. (Pure, or 24 K, gold is 1000 fine.) The fineness varies from stream to stream because of their different bedrock sources. However, in a watershed with a single source area, along with a course downstream for several miles the fineness will be found to gradually increase. This is because of the very slow solution of silver from the outer parts of the gold fragment. The smaller the gold particles, the more surface area an ounce of gold will have, and the greater the silver solution effect. Flour gold is purer than most nuggets. Older gold, as from a high bench or a fossil placer, will be purer than modern gold liberated from the same source.

ELUVIAL PLACERS

Placer gold is liberated from its bedrock source principally by weathering and occasionally by erosion. Weathering releases gold fragments in about their original shape and size. Fragments of gold released by erosion are reduced in size as the rock holding them is worn away. Weathering breaks up a rock by solution, by chemical processes such as hydration and oxidation that change mineral volume)and in that way fracture rocks) and by mechanical processes such as froze wedging and root wedging.

When weathering attacks a gold-bearing vein or other bedrock deposit, a portion of the vein minerals are removed in solution, and a few fragments may be removed by percolating ground water.

Because gold is highly resistant to chemical weathering, most of the gold fragments tend to remain in place, gradually moving downward toward less weathered rock as the mantle material is removed. A placer deposit concentrated in place by weathering processes is an eluvial placer. Since these placers are related to individual bedrock sources, eluvial placers are richest and most extensive where those bedrock sources are rich and extensive.

COLLUVIAL PLACERS

Eventually soil mantle material including weathered vein matter will be moved downslope by mass-wasting processes until it reaches the nearest stream course. Moisture in the soil mantle facilitates and hastens this movement. Flowing water moved principally over the surface of the soil mantle in sheet flood with no fixed course and in rill and gully wash. Lighter rock waste fragments are washed away at the surface but there is little downward concentration of the heavier minerals. Downslope movement of the soil mantle (with material from upslope sliding over material below), accentuated by processes such as frost heave, tends to lift heavy minerals away from bedrock.

ALLUVIAL PLACERS

An alluvial placer is one in which transportation, concentration and deposition of placer gravel by running water have been the dominant processes forming the deposit. These are the most important gold placers in Colorado. Alluvial placers may be beneath streams and their valley floors, beneath gravel terraces above the level of the valley floor or in alluvial fans, the cone shaped accumulations of debris formed where steep, mountain streams flow onto gently sloping lowlands. According to their settings, they are called valley or stream placers, terrace or bench placers and fan placers.

FAN PLACERS

Fan placers differ from the other types of alluvial placers. They are placers on alluvial fans, deposits of alluvium dumped by streams where they flow out of steep mountain canyons onto more gently sloping lowlands. The streams flow first along one course, then along others, depositing material and forming a cone shaped deposit. Placers may be formed on the upper part of the fan along several channels, rather than a single one. These channels usually do not reach bedrock; their floors are in fan material. Younger placers lie in material deposited after, and may lie over older placers, which is the contrary of other alluvial placers. Terrace No. 2 and the outwash plains below the moth of the canyon of Tarryall Creek in Park County have fan characteristics, but this fan has not been built up high at its head, and the conical shape is not readily apparent. The placer deposits of Buckeye Gulch in Lake County and Blank Gulch/Placer Creek in Chaffee county are on alluvial fans.

FOSSIL PLACERS

Fossil placers are placer deposits, originally of any type, that have become buried in the stratigraphic section beneath younger sedimentary rocks and/ or lavas. The placers that were worked in the drift mines at about 7,000 feet elevation in the Iron Springs Divide area in Moffat County are fossil placers, originally alluvial and perhaps marine deposits, contained in the Eocene Cathedral Bluffs member of the Wasatch formation. Fossil alluvial placers in the Oligocene Castle Rock formation at Cherry Creek Divide in Douglas Counties were the bedrock source of much of the gold of the Cherry Creek watershed.

WHERE TO PAN

RESPECT PRIVATE PROPERTY

Much of, but not all, Colorado's placer ground is private property. Just as when fishing or hunting, before panning on fenced or posted land, get permission! Do your shoveling and dispose of your tailings so they are inconspicuous, if not invisible! Before using a rocker or a portable dredge, get the land-owner's permission or if in a national forest a ranger's.

DANGERS

Although panning is not dangerous, there are a few special risks that you should avoid:

* Don't go into drift mines or other abandoned workings! All the drift mines have been abandoned at least 60 years and any abandoned working is hazardous. A mine needs continual maintenance to be safe.

* Be especially careful around streams more than knee deep The streams are deceptively swift, powerful and cold.

Rocks on the stream beds often are moss-covered and very slippery.

* Be careful around water-filled dredge ponds or other placer pits. They are often deep with sheer banks and the water is cold. Avoid the banks. If you must be near them, locate a place you could climb out should they collapse.

* Do not do scuba placering. Compared with the streams elsewhere where it has been done with more or less success, Colorado's mountain streams are too small and too steep for safety. Also many of these streams' attractive sections have already been cleaned to bedrock one or more times.

MAPS

The U. S. Geological Survey publishes the County Map Series at 1:50,000 scale, 30 x 60 minute series maps at l:100,000 scale (metric) and 7.5 Minute Quadrangle Series maps at 1:24,000 scale. All these maps show topography and name gulches, streams and other local topographic features. They are for sale at the U.S. Geologic Survey's Map Distribution Office in the Denver Federal Center Building 810, P. O. Box 25286, Denver, CO, 80225, as well as some local stores.

The U. S. Forest Service publishes maps of the National Forests and Surrounding areas that do not have as much detail as the U.S.G.S. maps, but are usually adequate, nevertheless. These maps are for sale at the U.S. Forest Service offices. 11177 11177 West 8th Ave., Lakewood, Co. 80225. and at Ranger Stations throughout the state.

GOOD AREAS FOR PANNING

The following placers are ones where you are likely to recover colors washing a few pans of gravel. They are also easy to locate.

ADAMS, CLEAR CREEK, GILPIN AND JEFFERSON COUNTIES

Clear Creek and Tributaries

Placers have been worked from place to place from Jackson's Bar on Chicago Creek, from the mouth of Fall River on Clear Creek, from the heads of Russell and Gregory Gulches west of North Clear Creek and from above timberline on North Clear Creek to Denver. This placering began the winter of 1859 and has continued with interruptions to the present. Jackson's Bar and Gregory Gulch were two of the three economic gold discoveries early in 1859 that sustained the Colorado gold rush. No commercial placer mining continues today.

Going up North Clear Creek from Forks Creek (at the junction of U.S. 6 and Colorado 119) there are signs of placer mining along the creek almost continuously to Pickle Gulch and in Russell and Gregory Gulches and many of their tributaries nearly to their heads. All have been placered from once to more than three times. The valley placers were worked by shoveling-in, where water was available (especially in the upper courses), by small scale by hydraulicking and along North Clear Creek from Black Hawk downstream by drift mining and by dragline and mobile dry land washing plants. Along North Clear Creek below Black Hawk you can see a few placer sites from place to place. A prominent one is in sec. 26 T. 3 S, R 72W on gravel terraces about 55 feet above the creek. There were eluvial placers along many veins in the district, and between them and the streams colluvial placers were worked in some places.

You can recover some colors fairly quickly at any of these sites. Along the main creek you cannot reach bedrock; in the channels there the valley gravels are 25 to 30 feet thick. Look for skim bars and for gravel on the downstream sides of large boulders. Of the gulches, Russell Gulch watershed and Gregory, Nevada, Bobtail and Packard Gulches are most promising but they are obstructed in places. On the terraces you can reach bedrock gravels in many places. The eluvial deposits are almost completely exhausted, but you might get a panful or two of good dirt from them. A good place to start would be climbing along the Gregory vein, marked by a line of pits and depressions that go uphill southwest from the Gregory monument beside the road between Black Hawk and Central City. You will see what a vein outcrop looks like and can then recognize others leading from some of the shafdts in the district. In places along the outcrops and underground workings are open. Do not enter them! Most have not been wroked for 80 years or more and are unsafe. The backs (roofs) and in many places even the floors are not secure.

Placers have been found from place to place from the mouth of Fall River downstream to Forks Creek and along Chicago Creek for about half a mile upstream from Clear Creek. The Chicago Creek section was "Jackson's Bar." The sire of Jackson's discovery was on the west side of Chicago Creek about a quarter of a mile south of Clear Creek. Between Fall River and the West end of Idaho Springs was "Spanish Bar."

Clear Creek heads in higher mountains than does North Clear Creek, in areas that underwent glaciation: whereas North Clear Creek's watershed did not suffer glaciation. The principal gold deposits lie in valleys that join Clear Creek below Fall River. (The gold de posits of North Empire and of Alice each gave rise to eluvial and colluvial placer deposits in tributary valleys, but glaciers and heavy outwash streams destroyed or buried any stream placers derived from them in the Clear Creek and Fall River valleys.) Consequently, there are no important placers on Clear Creek above Spanish Bar. Also because of the volume of the outwash streams, the valley placers from Spanish Bar downstream have thicker gravels and more and bigger boulders than do the placers on North Clear Creek.

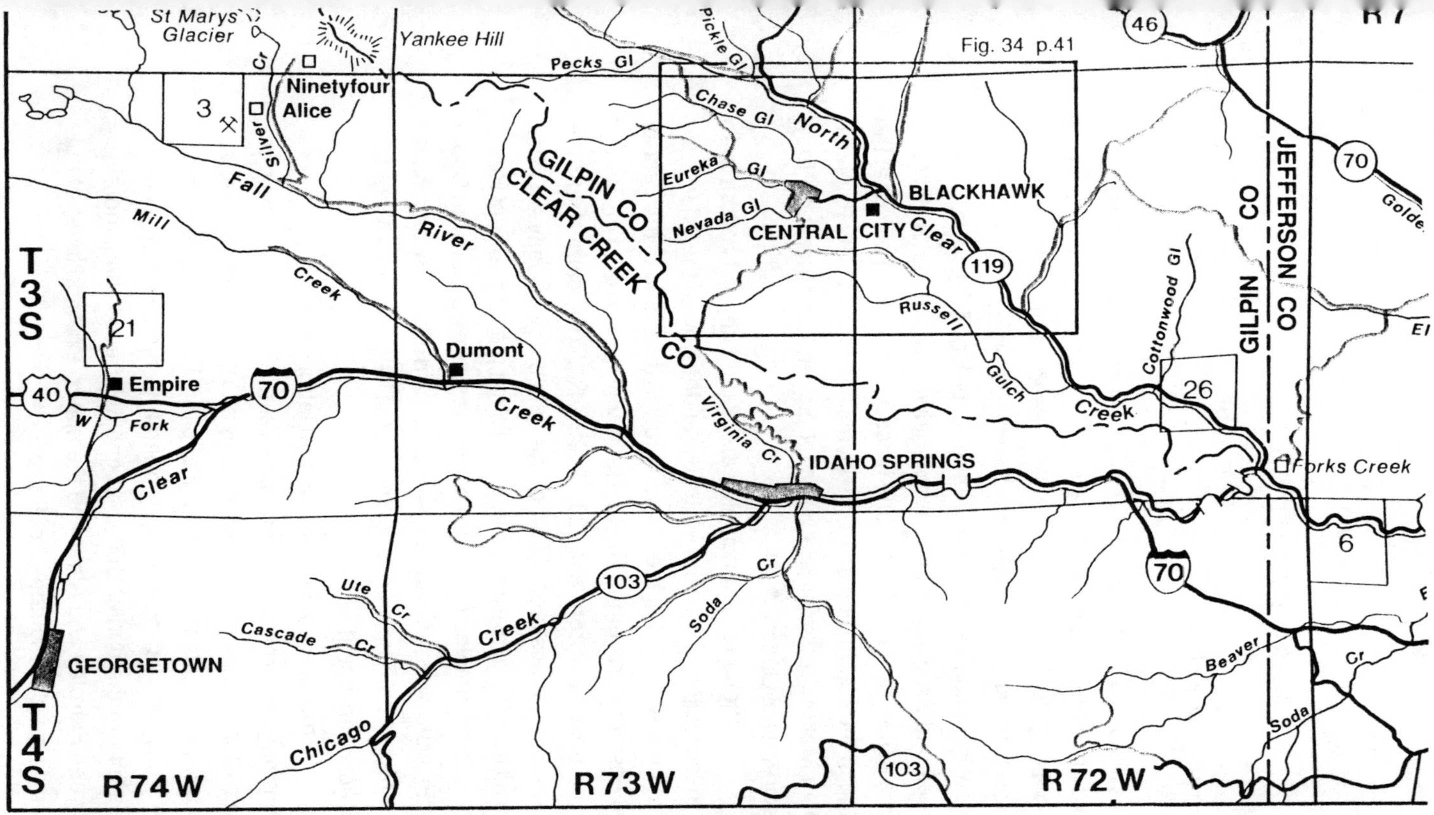

Figure 33. Index map of Clear Creek and tributaries placers, Adams, Clear Creek, Gilpin, and Jefferson Counties. (Modified from U.S.G.S. topographic maps)

PARK COUNTY

Fairplay district

This district in Park County has many placers. Although shows of gold have been found from the foot of Hoosier Pass downstream and along Buckskin and Mosquito Creeks to Alma, the principal placers are at Alma, along the South Platte River from Alma to below Fairplay, along lower Beaver Creek and between Beaver Creek and the South Platte about midway between Fairplay and Alma (Snowstorm or Timberline placer). An unusual placer at the edge of the district lies on Pennsylvania Mountain. The Pennsylvania Mountain placer is a colluvial placer, the others are from glacio-fluvial placers.

Although the principal bedrock gold deposits of the area lie to the west of the South Platte River, the flow of the Wisconsin glaciers was such that the ice transported their gold to the central and eastern parts of the South Platte Valley. The gold of Beaver Creek upstream from sec. 20, T. 9S., R. 77W. came from the Mount Silverheels area. Most of the gold of lower Beaver Creek and all the gold of the other Fairplay district placers, except Pennsylvania Mountain, came from the western part of the South Platte watershed--Mosquito and Buckskin Gulches and North Star Mountain. The source of the gold of Pennsylvania Mountain has not been identified.

From the earlier Illinoian glaciation (Bull Lake) there remains only the breached terminal moraine at Fairplay, some remnants of lateral moraine and extensive outwash buried beneath the outwash plain south-east of Fairplay. The Wisconsin glaciation (Pinedale) is marked by terminal moraine beside and upstream from the earlier Bull Lake one, two recessional moraines between Fairplay and Alma and the upper, exposed layer of outwash on the plain southeast of Fairplay.

There are remnants of terraces 25,55 and 180 feet above creek level near Idaho Springs and at a few places downstream to Forks Creek. Some placer gravel can be found at some of them and on the point bars and below boulders on Clear Creek.

From Forks Creek to the mouth of Clear Creek Canyon gravel piles beside the stream bed mark the site of dry land washing plant placer operations of the early 1930's. The banks beside some of them expose gravel banks beneath hillside mantle. There are a few terrace remnants in Clear Creek Canyon in sec. 6, T 4S, R.71W., and in sec 36, T.3S., R 71w., that correspond to the 55-foot terrace in Idaho Springs and North Clear Creek. These were worked at least in part in the 1930's. On the north slopes of Mount Zion (the mountain with Colorado Mines' "M") west of Golden, in secs. 32 and 33 T.3S., R.70 W there are old cuts in a terrace about 230 feet above the creek. These expose some good grade gravels. They probably weren't economic--it was as hard for the miners to bring gravel to the creek as it will be fore you if you attempt it.

From the mouth of the canyon downstream to the South Platte River, the gravels beneath Clear Creek valley are gold-bearing. In 1904 and 1905 attempts were made to work them with two dredges located in NE sec. 27, T.3S., R.70 W. This operation failed because there were many lava boulders too big to move on bedrock. Placer miners worked at the mouth of the canyon in 1892 and 1893 and in the early Depression years. Those working in the 1890's were reported to be making $3 to $5 per day, which would correspond to a production of one-sixth to one-quarter ounce per day. The Depression miners rarely made this much.

These miners did some drifting and also worked the skim bars. At the mouth of the canyon Clear Creek's gradient becomes more gentle and some gold can often be gotten from the top gravels at the heads of the point bars through Golden after the spring runoff and after heavy storms.

The sites in the canyon and the point bars in Golden are favorable sources of gravel for panning.

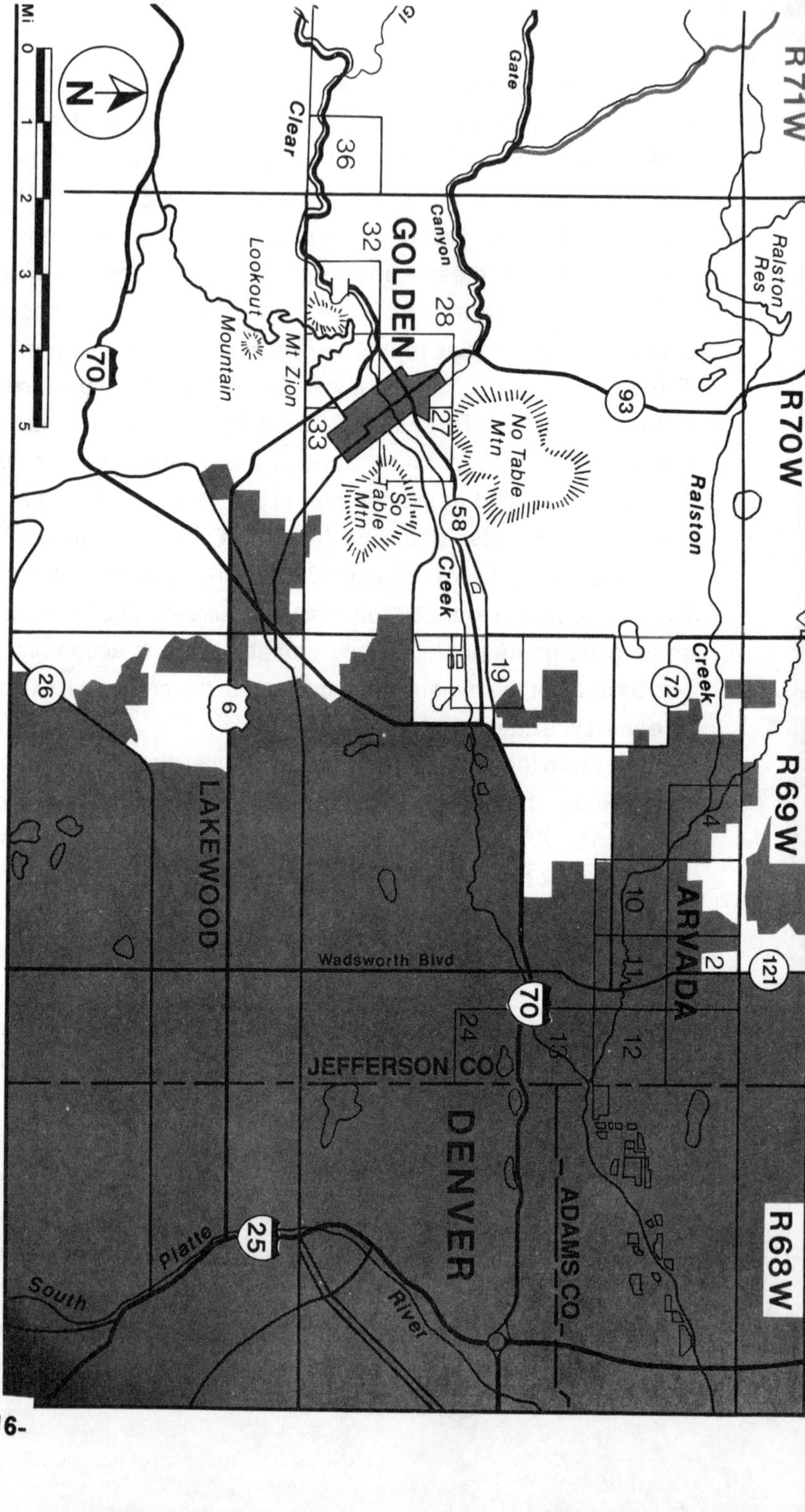
R71W
R70W
R69W
R68W
Ralston Res
Ralston
Creek
ARVADA
Gate
Canyon
GOLDEN
No Table Mtn
So Table Mtn
Creek
Clear
Mt Zion
Lookout Mountain
Wadsworth Blvd
JEFFERSON CO
ADAMS CO
DENVER
LAKEWOOD
River
Platte
South
121
93
72
58
70
6
25
26
2
4
10
11
12
13
19
24
27
28
32
33
36
N
Mi 0 1 2 3 4 5

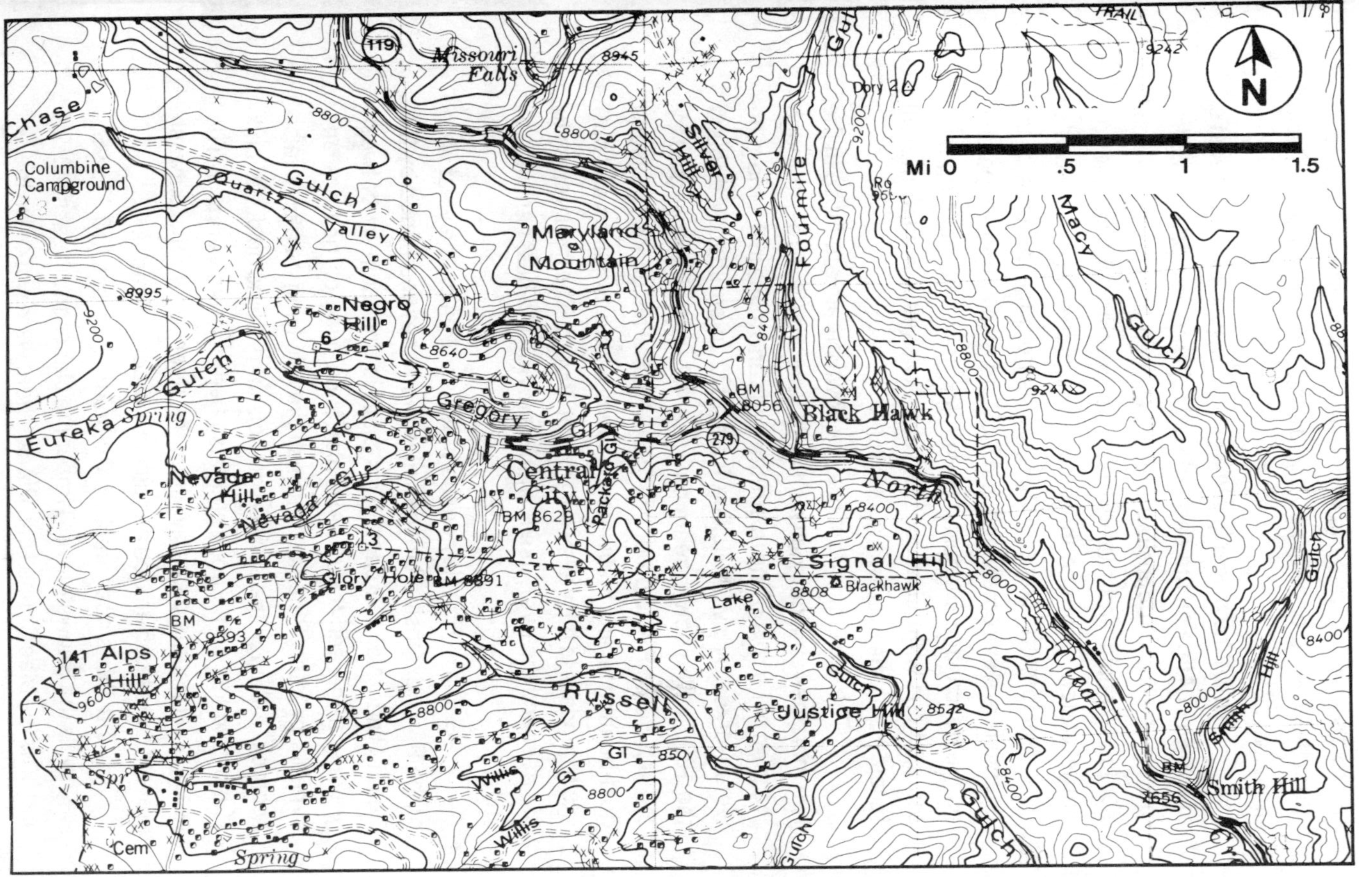

Figure 34. *Topographic map of Central City and Black Hawk area, Gilpin County. (U.S.G.S. County Map Series)*

(A recessional moraine is a moraine that is formed upstream from the lowest, or terminal, one during a minor surge of glacial activity.)

The Alma placer lies in a hill to the east and south-east of the town of Alma on the east side of the river. The placer is composed of alluvial deposits formed between the last two glacial surges overlain and mixed with gold bearing till from the last surge's lateral moraine in which some glacio-fluvial meltwater channels formed. Some of the gravels from the bottoms of the channels are extremely rich, but they are also extremely patchy, as the description suggests. Here, pick gravel for panning from the sections with most cobbles, and be careful around the cut faces; from time to time gravel falls!

The next important placers downstream are those in the two moraines and their outwash from Sacramento Creek through the town of Fairplay. The upstream parts and those in the moraine were worked by hydraulicking from 1872 intermittently through the 1890's. The downstream part near river level was worked by the first South Platte dredge in 1922-1925. This deposit includes gold-bearing till and meltwater channels in the moraines and outwash deposits in the valley cut through thyem. The banks close to the river and the river, itself, should provide the best panning results.

South of Fairplay on the outwash plain on the west wide of the River extensive boulder piles mark the area mined by the South Platte No. 1 dredge. This plain is underlain by outwash gravels of the early, Bull Lake glaciation lying on bedrock and overlain by later, Pinedale outwash gravels lying on a clay false bedrock. Although the surface gravels carry some gold, the bulk of the gold lies on bedrock and false bedrock, out of reach. The best gravel here is probably along the river, where the outwash gravels have been eroded and sllightly reconcentrated.

On the slopes east of Colorado 9 in sec. 19, T. 9S., R.77W. is the Snowstorm or Timberline placer and in secs. 20, 29 and 30 is the Mac Connell cut. Placers here were worked in the

1870's, the Snowstorm was hydraulicked in 1902 and 1903 and the MacConnell cut worked from1904 to 1916. The Timberline dragline floating washing plant was installed in 1941 and worked in 1942, 1946 and 1947. These placers are deposits in meltwater channels in Bull Lake and Pinedale Wisconsin lateral moraines. These moraines contain masses of alluvial placer gravels formed in interglacial times, mixed with till and pushed onto the hillside by the glaciers. Gold was concentrated on Bedrock, but in the southern part of the placer there were two false bedrock streaks, as well. For panning select the lowest and coarsest gravels to be found in the old workings.

Along Beaver Creek upstream to the southeast corner of sec. 17 and from there to the northwest in Poorman Gulch for a short distance are placers formed chiefly by meltwater streams flowing along the edge of Pinedale moraines. These placers were of good grade, profitably worked. Upstream from the southeast corner of sec. 17 there are distinctly poorer stream placers to the slopes of Mount Silverheels.

PENNSYLVANIA MOUNTAIN

The Pennsylvania Mountain placer is on the upper slopes of Pennsylvania Mountain in secs. 16, 17 and 20, T. 9S., R.78W. at an elevation of about 12,250 feet. It lies between Pennsylvania Gulch and the cliffs overlooking Mosquito Creek on an unglaciated slope that has a bedrock trough extending from near the center of sec. 17 to near the center of sec. 16. Most of the old placer workings are along this trough. On the slope there is up to 19 feet of mantle material composed of angular quartzite and rounded porphyry cobbles and boulders in a sand matrix.

The gold is somewhat flattened but still has sharp edges and often shows crystal faces. The placer is well known for the coarse gold that has been produced there. The largest nugget produced weighed 11.95 ounces--"a pennyweight less than a pound." In 1990 Shane Dodge found an 8.6 ounce nugget and a 12 ounce one was found in 1937. There have been many one-ounce samples. The gold produced is notable for the low proportion of fine gold that it contains. The gold is not concentrated on bedrock but rather in the central part of the gravel section. This is a colluvial placer. The bedrock source of the gold has not been identified.

A great problem at this placer has been the scarcityc of water. At one time the gravel was carried by aerial tramway from the placer to Mosquito Creek. Other times water has been caught in dams on Pennsylvania Creek and recirculated. Neither solution was very effective. Should you prospect here, look for gravel about midway between the gulch on the Mountain and the cliffs in an area where there are many old pits. Remember that with such coarse gold, good grade gravel will contain only a few fragments in a cubic yard; There will be many barren pans. *The placer area is private property--get permission before you go there!*

TARRYALL CREEK AREA

Gold was discovered near the forks of Tarryall Creek in late summer of 1859 and worked, usually on a small scale, with some interruptions until the 1890's. Many individuals placered here during the Depression. A dry land plant was set up on the Wilson placer in 1934 and worked until 1942 and in 1941 to 1942 and 1946 to 1947 a dragline dredge was operated on Tarryall Creek below Peabody's and on Cline Bench. At about the same time there was extensive work in Little French Gulch. In the 1980's there was again some activity in the valley.

The important placers can be considered in three groups: 1) Those in the canyon of Tarryall Creek (Fortune and Peabody placers), Middle Tarryall Creek, Little French gulch, the lower mile of North Tarryall Creek, and smaller placers in Deadwood and Montgomery Gulches. 2) Those on Tarryall Creek in South Park. The valley floor of Tarryall Creek from the mouth of the canyon downstream for about a mile in secs. 19 and 20 and a bench

bench placer in secs. 21, 22 and 27, T. 8S., R. 76 (Cline Bench placer. 3) The placers on North Park Gulch in sec. 1, T. 9S., R76 W in Park Gulch in sec. 8 and 17, T. 9S., R 75. (Wilson Placer) and downstream on Tarryall Creek in sec. 10.

The bedrock source of these placers is myriad tiny gold veins surrounding an igneous stock on the north slope of Mount Silver Heels and extending from Montgomery to Deadwood Gulches. Although none of these veins have proved economic to mine, in the aggregate they have provided much gold for placer concentration.

In the extreme heads of Montgomery, Little French and Deadwood Gulches are colluvial placers that in a few spots have been of sufficient grade for limited placering. These are unusual because they have developed since the final disappearance of the glaciers; before then they were beneath the ice. They grade downslope into glacio-fluvial and glacial placers.

The Wisconsin Pinedale terminal moraine is at the narrows in the valley in sec. 18, T. 8 S,., R 76 W., and sec 13, T. 8S., R 77 W. Above this terminal moraine there are recessional moraines about at the mid-point of Middle Tarryall Creek, at the mouths of Montgomery and Little French Gulches and further upstream in Montgomery and Deadwood Gulches. The terminal moraine has been worked extensively in lower Tarryall and North Tarryall Creeks and the slope between them at the Fortune placer. The workings develop meltwater channels across and along the terminal moraine, outwash from the recessional moraine upstream and apparently some glacial till unwashed by later streams. This moraine is unusual in that it is all slightly gold-bearing and some unwashed spots have been found that paid to placer. It is also interesting that the gold in North Tarryall Creek came from Middle Tarryall Creek. In secs. 11, 12, and 13 the Middle Tarryall Creek Pinedale glacier flowed northeastward over the divide between the two valleys. There is a remnant of Illinoian Bull Lake moraine below the narrows to the edge of South Park in

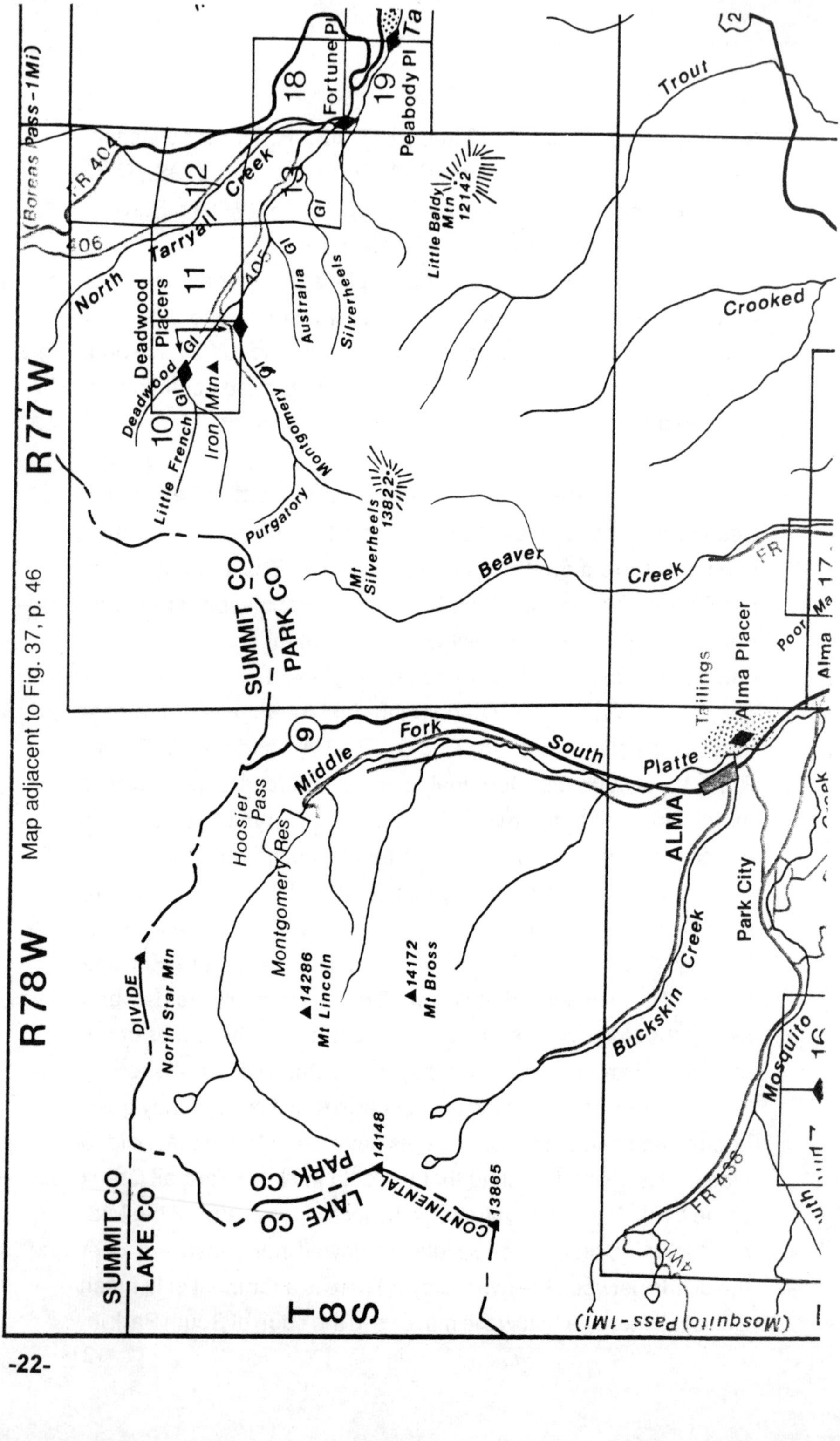

Map adjacent to Fig. 37, p. 46

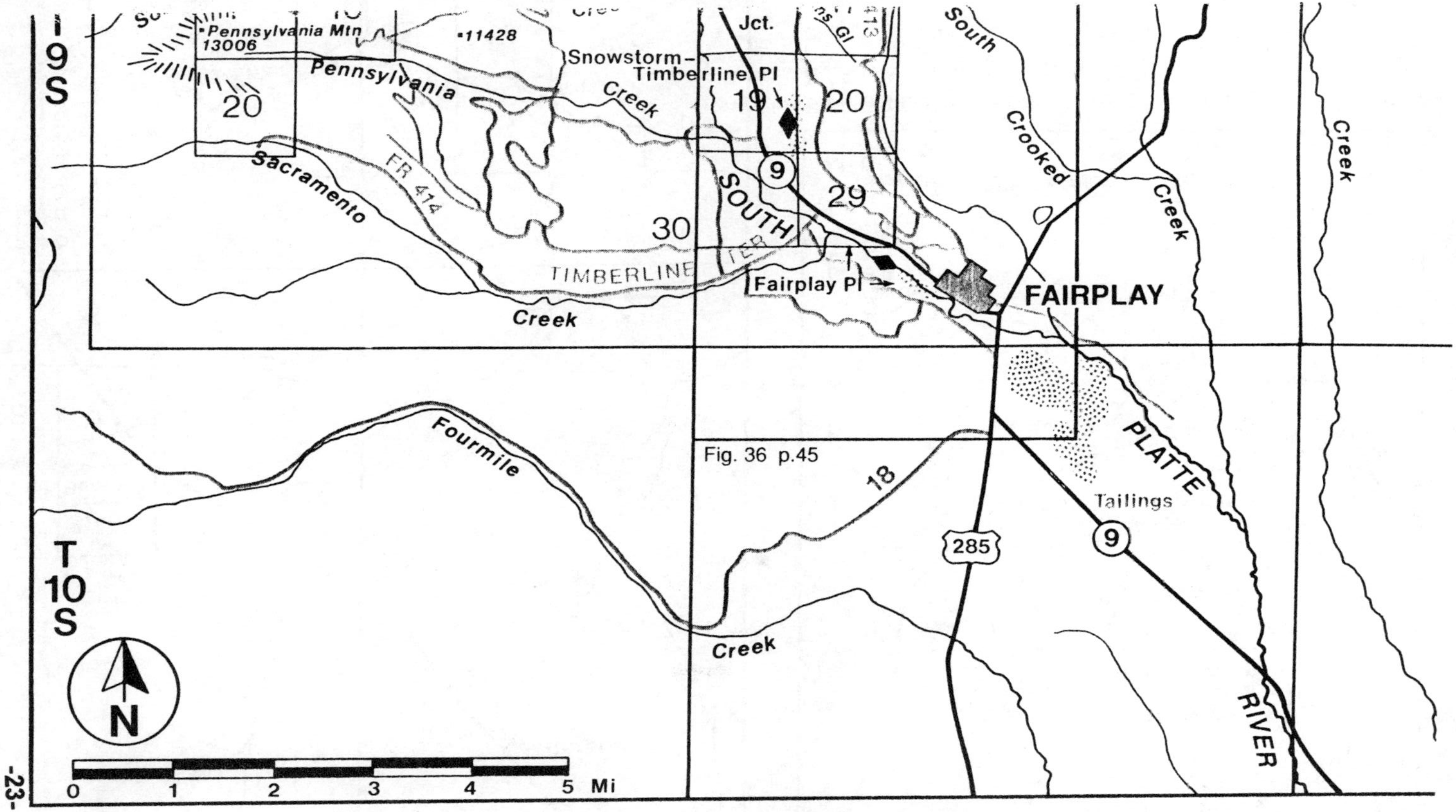

Figure 35. Index map of Fairplay–Tarryall district placers, Park County. (Modified from U.S.G.S. topographic maps)

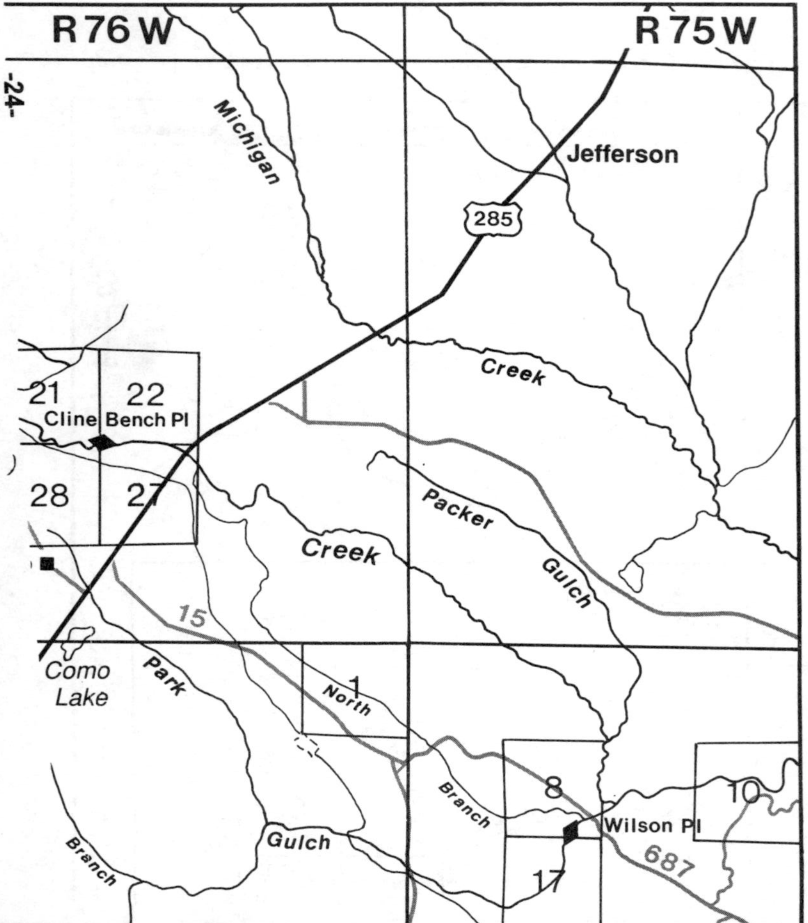

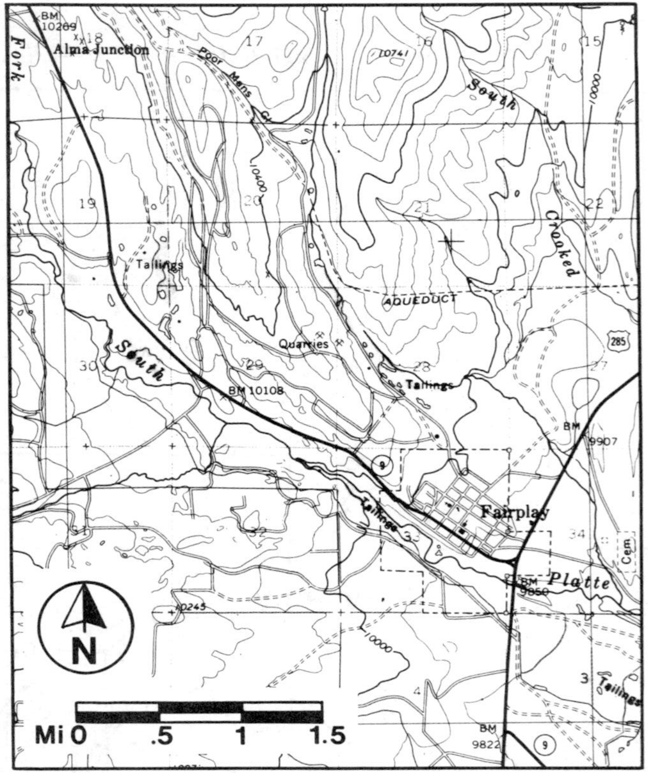

Figure 36. Topographic map of Fairplay, Park County. (U.S.G.S. County Map Series)

in sec. 19 and into 20, T. 8 S., R. 76 W. In sec. 19 where Tarryall Creek flows on the north edge of the Bull Lake moraine, all the valley floor has been placered (Peabody placer). These deposits were outwash from the Pinedale glacier and reworked Bull Lake till.

From the mouth of the canyon eastward there are terraces at three levels above the floor of Tarryall Creek. The two higher terraces lie on the south side of Tarryall Creek in secs. 27 and 28 and the lower is on the north side of the creek there. The higher terrace (called Terrace No. 2) was formed by Tarryall Creek in pre-Bull Lake time. The terrace below it is the Bull Lake outwash plain. The bottom terrace is the Pinedale outwash plain, into which Tarryall Creek has cut its modern valley floor. Each of these terraces is more or less gold bearing for a few miles away from the canyon.

The valley floor of Tarryall Creek, worked for more than a mile below Peabody's was a Pinedale glacio-fluvial deposit enriched by post-glacial concentration. The Cline Bench deposit was a Bull Lake glacio-fluvial deposit in a bedrock stream channel in outwash. There may be other channels like these beneath each terrace.

The immediate source of the placers on North Park Gulch and in Park Gulch (Wilson placer) is the gravels of Terrace No. 2 and the bedrock source is the veins at the head of Middle Tarryall Creek. The placers downstream on lower Tarryall Creek probably have some Bull Lake and Pinedale outwash gold as well.

In the stream placers, select gravel for panning from as low as in the faces as possible. In Cline Bench and adjoining it on the south side of Tarryall Creek, bedrock may be exposed from place to place. In the placers in the canyon, very coarse or cobbley sections in the cuts may be meltwater channel floors and may be well worth panning. At the Fortune placer, since the till is somewhat gold-bearing, sample the moraine banks from place to place, as well. At the heads of the valleys old pits may suggest the

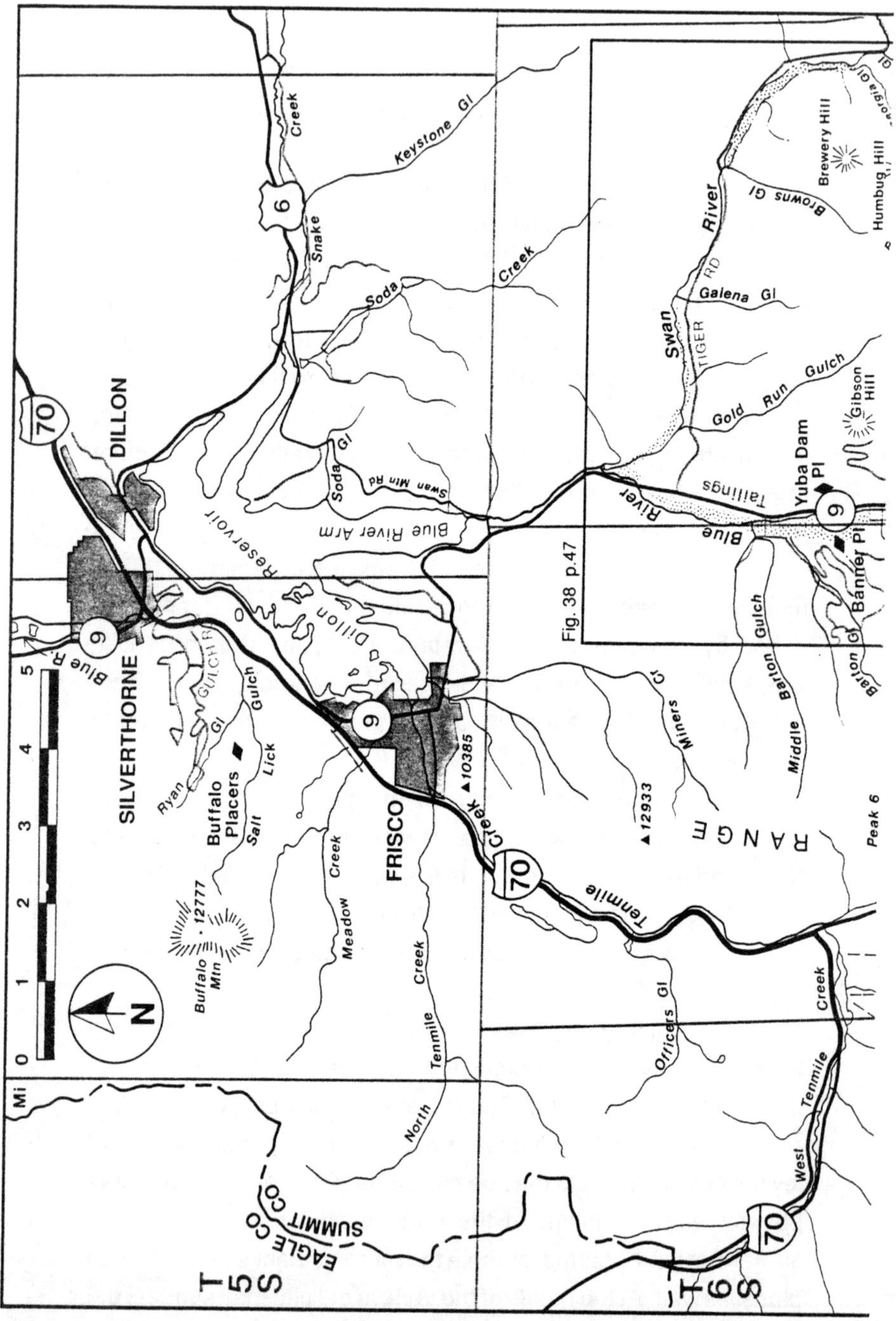

Mi
0 1 2 3 4 5
N
SILVERTHORNE
DILLON
FRISCO
Blue R
GULCH RD
Ryan Gl
Buffalo Placers
Salt Lick Gulch
Buffalo Mtn
12777
Meadow Creek
North Tenmile Creek
Tenmile Creek
West Tenmile Creek
Officers Gl
Miners Cr
10385
12933
RANGE
Peak 6
Dillon Reservoir
Blue River Arm
Soda Gl
Swan Mtn Rd
Soda Creek
Snake Creek
Keystone Gl
Fig. 38 p.47
Swan River
TIGER RD
Galena Gl
Gold Run Gulch
Browns Gl
Brewery Hill
Humbug Hill
Gibson Hill
Yuba Dam Pl
Banner Pl
Blue River
Tailings
Middle Barton Gulch
Barton Gl
EAGLE CO
SUMMIT CO
T 5 S
T 6 S
70
9
6

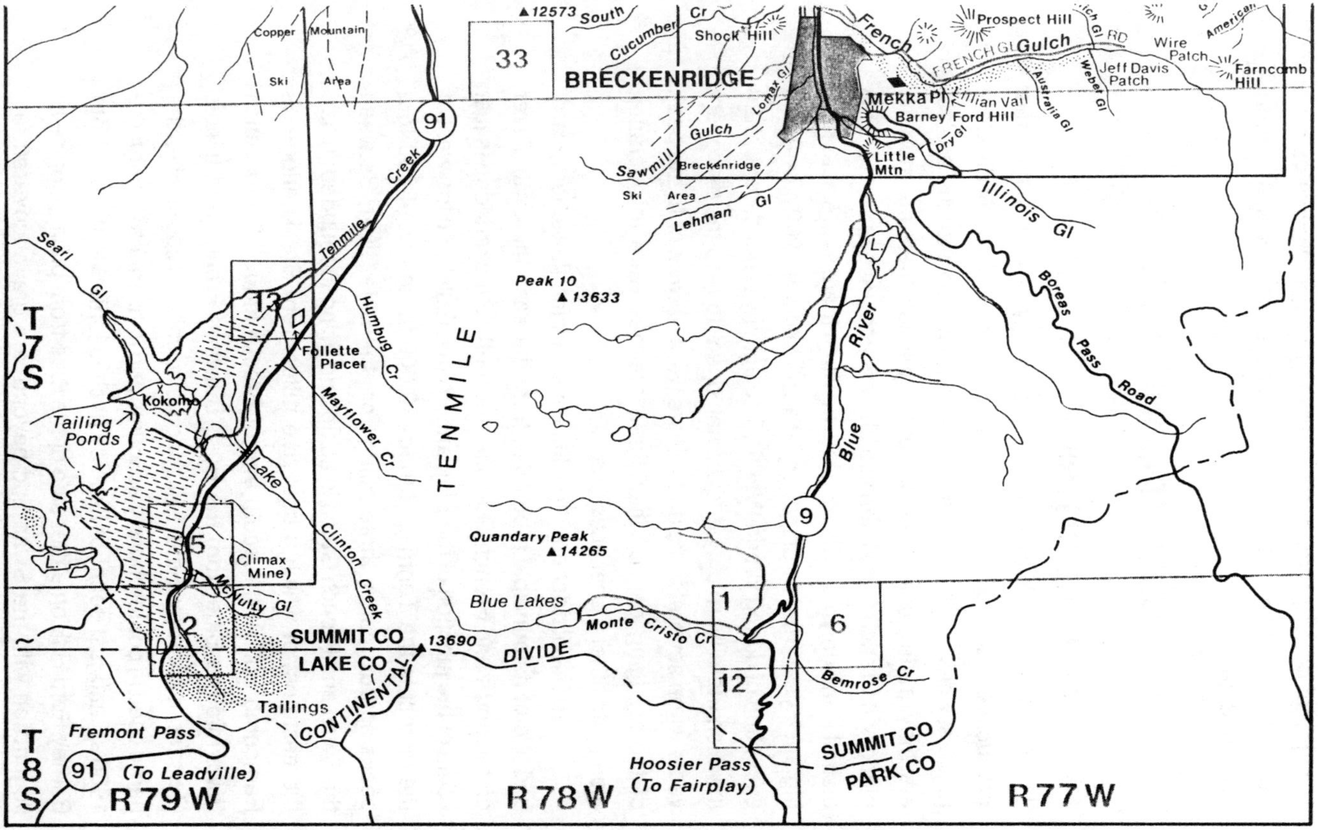

Figure 37. Index map of Breckenridge district placers, Summit County. (Modified from U.S.G.S. topographic maps)

location of better grade spots. The gravels contain lots of magnetite in black sand grains and pebbles and cobbles and abundant epidote and garnet.

SUMMIT COUNTY

BRECKENRIDGE DISTRICT

The Breckenridge district, in Summit County, has been the most productive placer district in the State. It includes many different placers, which will be discussed by type. There are colluvial placers; they grade into al-luvial placers in gulches; there are moraine placers in French Gulch, terrace placers and valley placers. This district is the onlyh one in the State in which there was placer mining each year from its discovery in 1859 into the 1960's (with the exception of during World War II). The activity ranged from years in which there were only a few small operations to the discovery years when several thousand miners were at work or to the peak of the dredging period when three dredges were in operation along with other placer activities.

In the district are the Blue River, which flows northward, and two of its principal tributaries from the east, the Swan River on the north and French Gulch to the south. Breckenridge is just south of the junction of the Blue and French Gulch. To the west of the district is the Tenmile Range--12,000 to nearly 14,000 foot peaks approximately three miles from the Blue River. Between the Swan and French Gulch is an upland rising to rounded "hills", the eastmost of which, about five miles from Breckenridge, is Farncomb Hill. At the south end of Breckenridge, Illinois Gulch flows into the Blue from the East. Between French and Illinois Gulches is Barney Ford Hill (originally called Nigger Hill, the name used in contemporary accounts and in histories through the 1960's), and just south of Illinois Gulch is Little Mountain. From Barney Ford Hill, on either side of the Blue from place to place to its junction with the Swan and beyond, are gravel-covered terraces usually 200 to 250 feet above the River. Small gulches tributary to these main streams drain the "hills" and the terraces.

Among these, Georgia and American Gulches and the Wire Patch on Farncomb HIll were noteworthy for their exceptionally rich placers. Georgia Gulch shares with California Gulch in Lake County the reputation of being the richest and most productive placers (for their size) in the State.

The high peaks of the Tenmile Range are relics of the pre-Summit surface mountains. The summits of Farncomb Hill and the other "hills" between the Swan and the Blue are remnants of the sub-Summit group of surfaces. The high terraces are older than the Illinoian Bull Lake glaciation. Terminal moraines of the upper Blue River glaciers just reach Barney Ford Hill, those of the French Gulch glaciers are about three miles east of Breckenridge, and those of the upper French Gulch (at Jeff Davis Patch and Rich and Weber Gulches) and the extremem lower parts of Georgia and American Gulches, the placered areas either were not glaciated or were downstream from the terminal moraines. Thye gravels worked by the dredges beneath the valley floors were outwash somewhat reworked by post-glacial streams. The gold in the valley gravels was concentrated close to bedrock; there were no extensive gfalse bedrock levels. Apparently here, in contrast to the Fairplay placer, the Bull Lake outwash was largely reworked by the Pinedale outwash streams.

There were numerous sources and a variety of kinds of sources of the placer gold. The source of the gold of Georgia and American Gulches and of the Wire Patch (and a part of the gold in the terraces and major streams) were the veins of Farncomb and nearby Humbug and Brewery Hills. These were many tiny veins--extending hundreds of feet, typically one-half inch thick and nearly barren. They were much explored. Miners might drift on a vbein through lond distances of barren ground and, if lucky, find a spot an inch wide and two or three feet in diameter filled with hackly wire gold. One day would repay a year's work. There are beautiful specimens of this gold on display at the Denver

Museum of Natural History and at the Colorado School of Mines library in Golden.

Veins of this kind, since eroded away, and small stockworks--intensely fractured zones whose many cracks are mineralized--are thought to have furnished the gold of the many hillslope (colluvial) and gulch placers from Galena Gulch to Gold Run and to the Swan River below them. Veins on Barney Ford Hill and Little Mountain supplied the gold to the rich Lillian Vale and Dry Gulch placers in gulches on the east end of Barney Ford Hill and to the terrace and stream placers downstream. Veins in the Dakota Sandstone in Shock Hill, just west of Breckenridge, and in Little Mountain contributed gold to the terrace and stream placers. Finally, some gold must have been contributed to the terrace and stream placers from deposits in the high mountains at the heads of the Blue River and Illinois Gulch.

Nearly all the deposits except those of the last type lay within 300 or 400 feet of the sub-Summit surface. The importance of these deposits as sources is related to their being close to this surface and to the long period of weathering to which the surface was subjected. Gold in the original deposits slowly dissolved in groundwater, moved downward to the water table and reprecipitated in coarser masses that were eventually liberated in the canyon cutting cycle. Gold deposited in this manner is called supergene gold. It is thought that the placers of the area between Gold Run and Galena Gulch lie below the level of the supergene zone in Farncomb Hill--type veins now destroyed--veins that probably supplied the placers' gold.

There were extensive colluvial placers beside and between Galena Gulch and Gold Run, at the heads of Georgia and American Gulches and at the Wire Patch. These deposits, up to 40 feet thick are of angular porphyry cobbles and boulders in a clayey matrix. Those in Georgia and American Gulches and the Wire Patch are nearly worked out. There are still some colluvial gravels on the west wide of Gold Run and between it and Galena

Gulch. The latter are probably better grade.

The gulch alluvial placer material is very similar to that of the colluvial placers--angular cobbles and boulders in a more or less clayey matrix. Composition of the course material varies with the bedrock in the gulch--porphyry or quartzite or both. These placers are confined to the gulches. Nearly every gulch flowing off the "hills" or the terraces has been placered to some degree. Any of the gulches may yield good panning gravel. Gravels as close to bedrock as possible should be the highest grade of any now exposed in cuts. Gulch gravels were hydraulicked in Galena Gulch. The bedrock gravel at the head of the hydraulic cut is reported to be good. The lower end of the hudraulic cut was not worked to bedrock. It may be possible to find good panning gravel here beneath recent wash and material caved in.

The terrace gravels were worked in many places. The most extensive placers were the Mekka placers on the north slope of Barney Ford Hill, the Yuba Dam placers on the east side of the Blue one mile north of Breckenridge, the Lomax Gulch and the Banner placers on the west side of the Blue (the first between Sawmill Gulch and Shock Hill, the second between Cucumber and South Barton Gulches) and the Buffalo placers on Ryan and Saltlick Gulches west of the Dillon Reservoir. The gravels very from 10 feet thick at the Mekka placer to more than 105 feet thick at the Buffalo placers. There were several discontinuous false bedrock levels.

Select panning gravel off bedrock where it is exposed; where it is not select gravel from the most cobbley sections of the banks. The south end of the Mekka placer cut reportedly still has some good grade gravels.

In the moraine placers in French Gulch at Rich and Weber Gulches select panning gravel from any sorted, cobbley zones (these might be meltwater channels). If these cannot be found, select gravel from places in the banks that have more abundant round cobbles and boulders for these may be places where a greater proportion of stream gravel is mixed with a

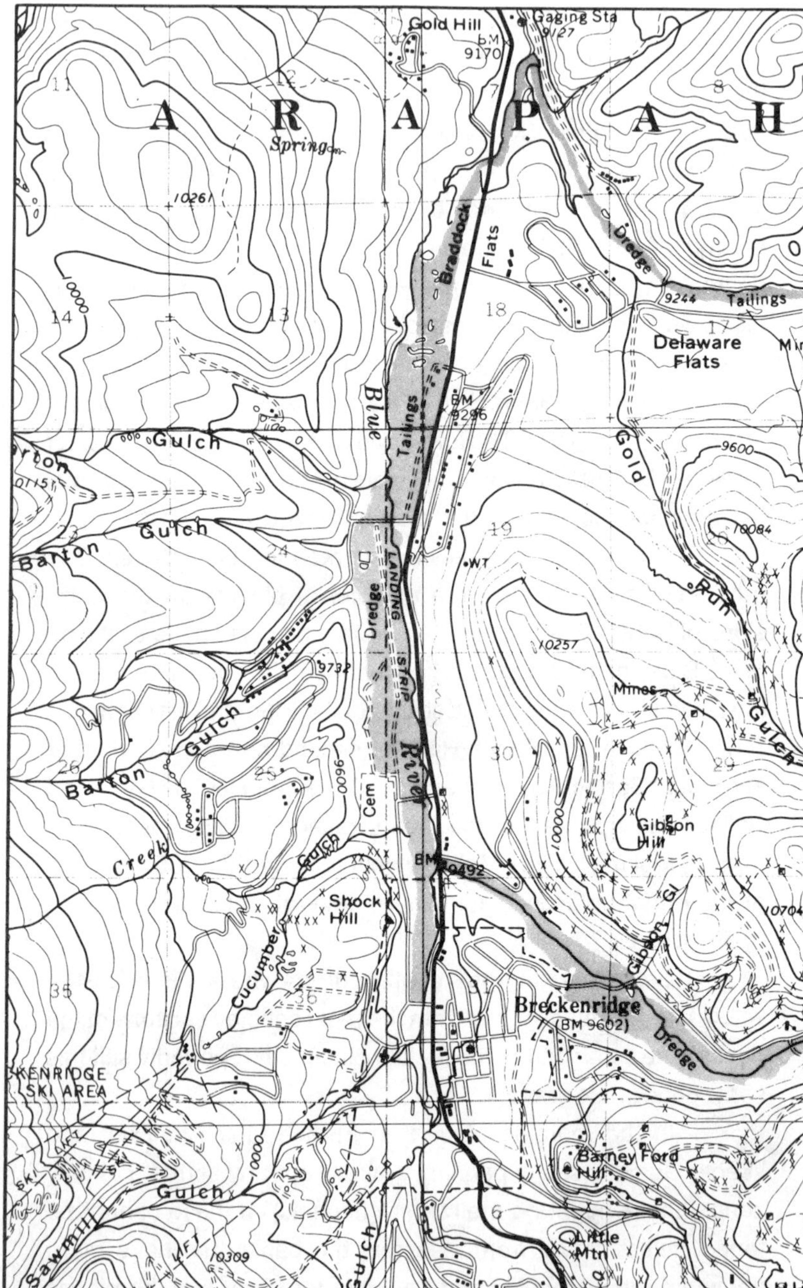

Figure 38. Topographic map of the Breckenridge area, Summit County.

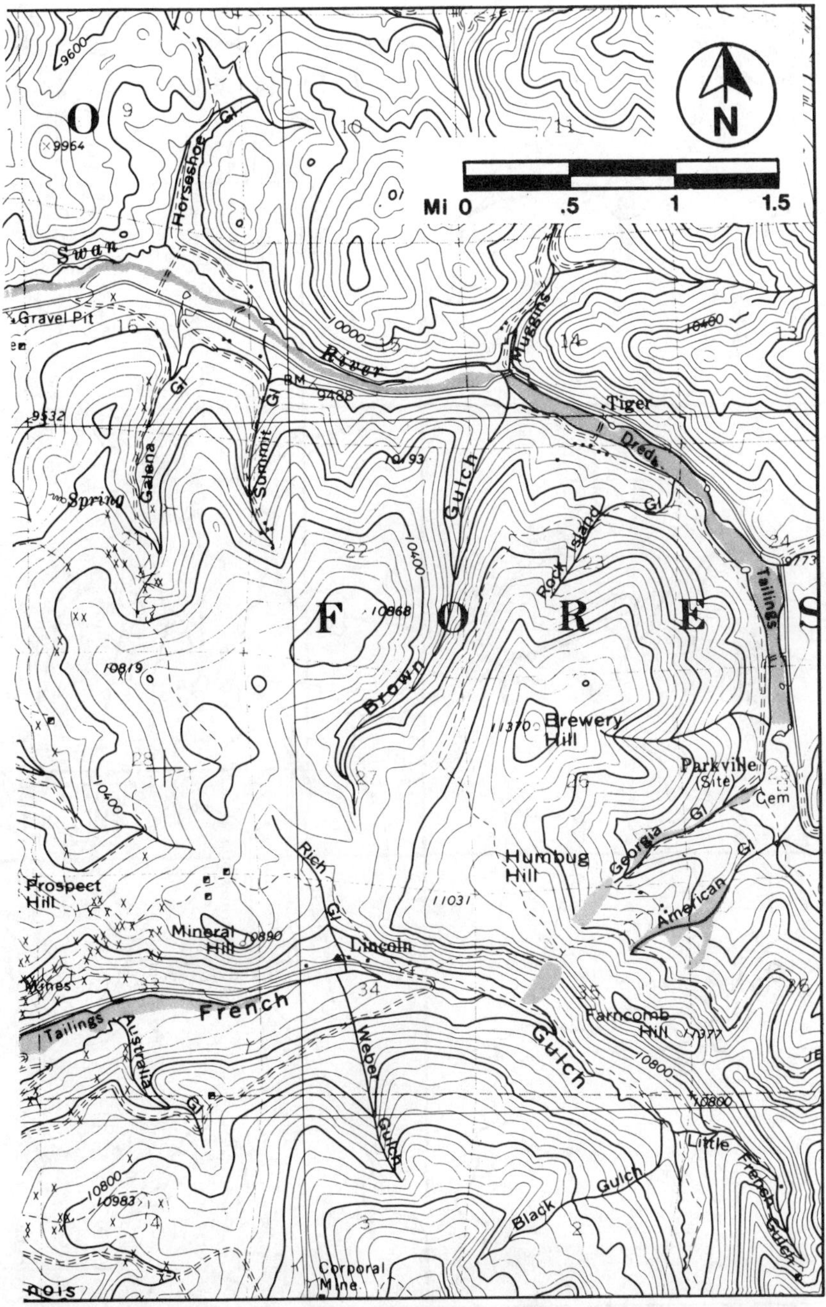

(U.S.G.S. County Map Series)

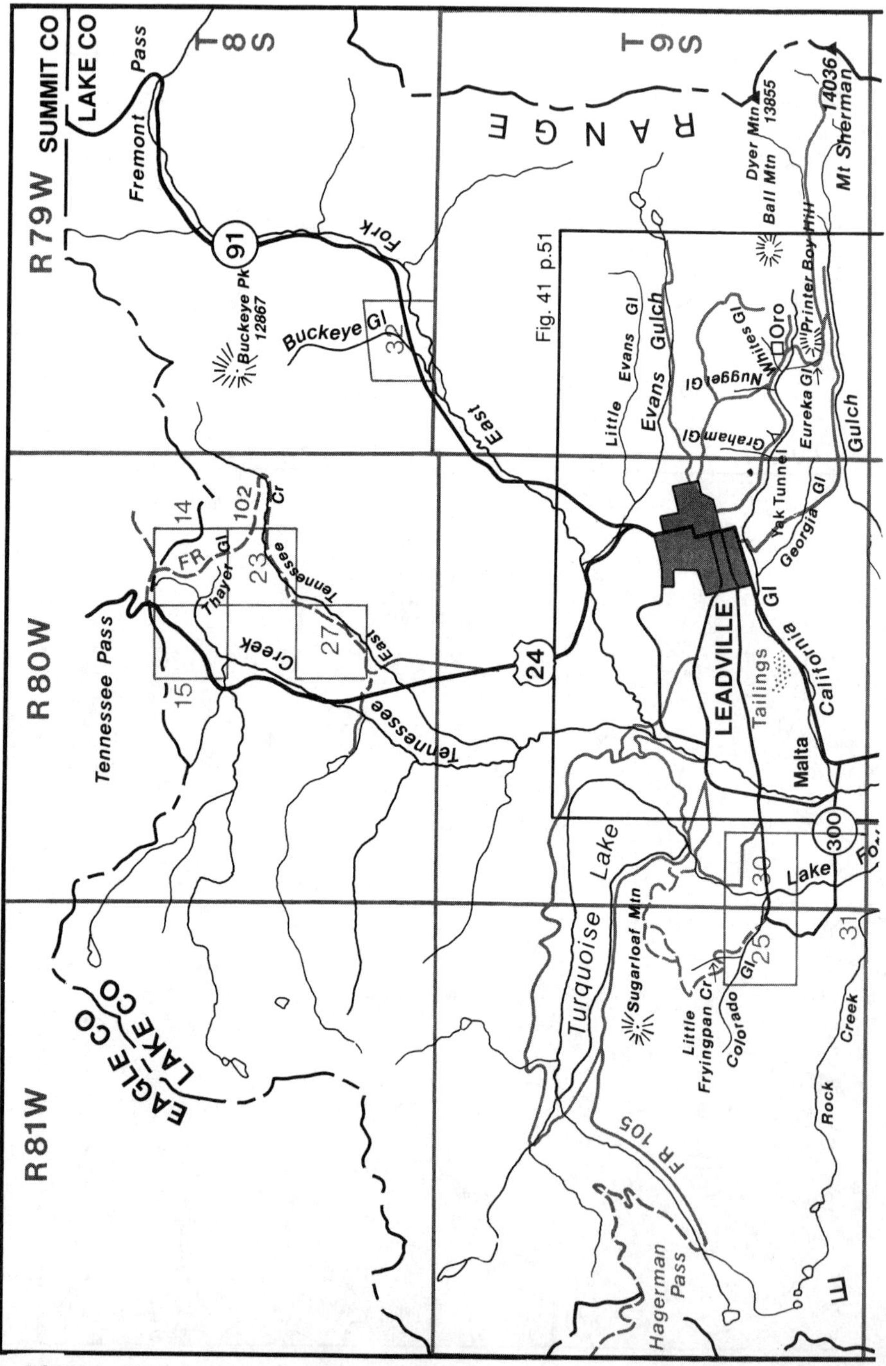
R81W
R80W
R79W
SUMMIT CO
LAKE CO
EAGLE CO
LAKE CO
T 8 S
T 9 S
RANGE
Fremont Pass
Tennessee Pass
Hagerman Pass
91
24
300
Buckeye Pk 12867
Buckeye Gl
East Fork
Tennessee Creek
East Tennessee Cr
Thayer Gl
FR 102
FR 105
Fig. 41 p.51
Little Evans Gl
Evans Gulch
Nugget Gl
Whites Gl
Graham Gl
Oro
Printer Boy Hill
Eureka Gl
Yak Tunnel
Georgia Gl
California Gl
Dyer Mtn 13855
Ball Mtn
14036 Mt Sherman
LEADVILLE
Tailings
Malta
Turquoise Lake
Lake Fork
Sugarloaf Mtn
Little Fryingpan Cr
Colorado Gl
Rock Creek
14
15
23
27
32
30
25
31

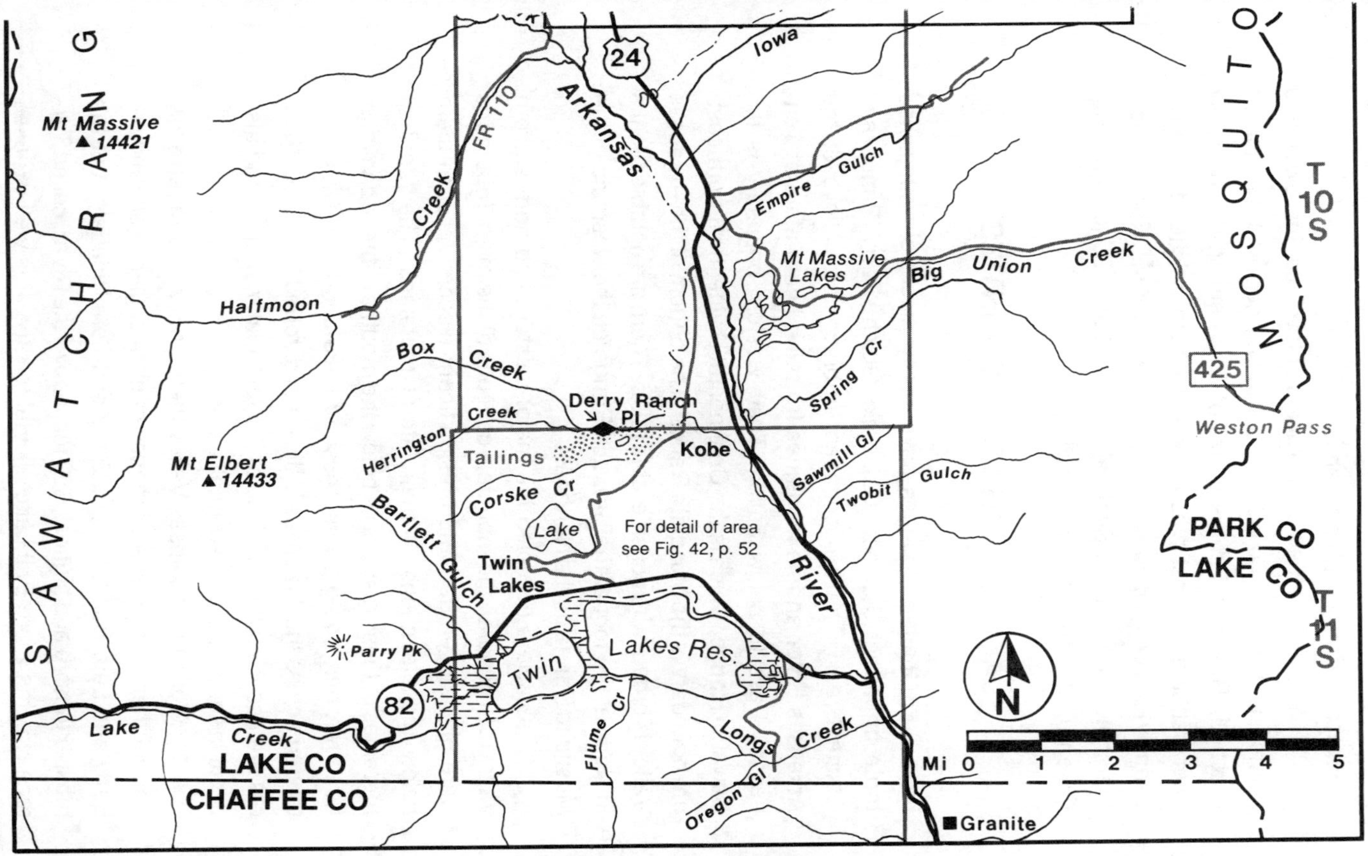

Figure 40. Index map of Lake and Chaffee County placers. (Modified from U.S.G.S. topographic maps)

smaller proportion of glacial debris

The gold in the valley placers of the Blue, the Swan and French Gulches is concentrated on bedrock and cannot be reached. The gravel in the banks of these streams is probably poor panning at most places except where gulches that have been extensively placered flow into these streams. For instance, the south banks of the Swan at and just below its junction with Galena Gulch are worth prospecting.

LAKE AND CHAFFEE COUNTIES

Lake and Chaffee Counties have many placer showings and three important deposits--California Gulch, Derry Ranch and Cache Creek. The two counties embrace the valleys of the Arkansas River and its tributaries from the headwaters at Tennessee Pass to below Salida. Their western boundary is the Sawatch Range, which is the Continental Divide; their northeastern boundary is the Mosquito Range. The bottom of the Arkansas Valley is broad, with considerable relief. There are notable breaks in slope at the foot of the Sawatch Range on the west and of the Mosquito Range on the Northeast.

Glaciers occupied the major western and northeastern valleys. In general, the mountain canyons of the major tributaries were glacially eroded and no significant placers occur in them below the elevations of their lateral moraines. The terminal moraines of these glaciers lie below the mouths of these canyons; however, only two glaciers, those of Lake and Clear Creek extended across the Arkansas Valley and diverted the River. Placers are associated with some of the terminal moraines and also occur in some unglaciated small valleys.

Along the Arkansas Valley there are extensive terraces. Four of them are outwash terraces of distinct glacial advances--Early and Late Bull Lake and Early and Late Pinedale--but the most extensive and highest, 360 feet above the River, called Terrace No. 1 is a pre-Bull Lake alluvial terrace. It is well exposed

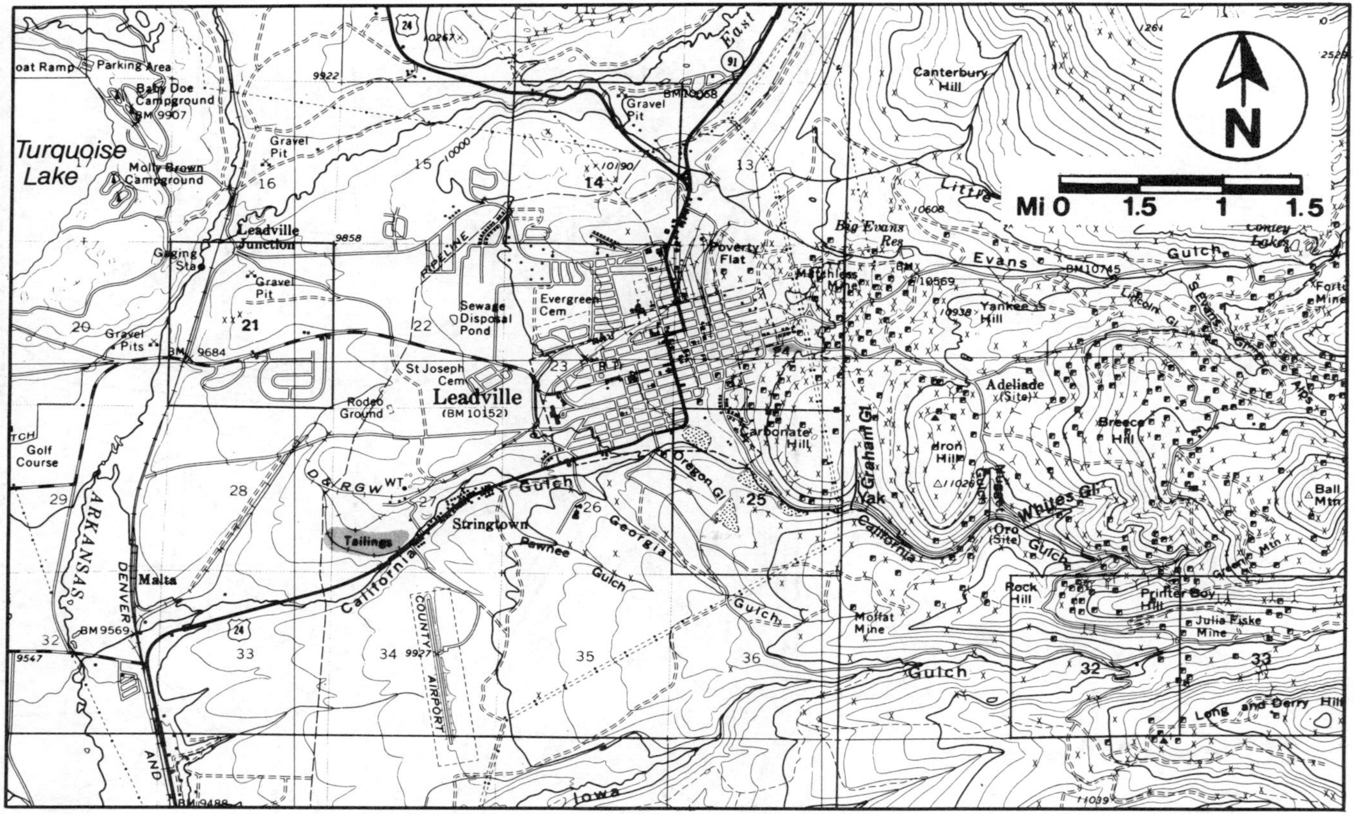

Figure 41. Topographic map of the Leadville area, Lake County. (U.S.G.S. County map Series)

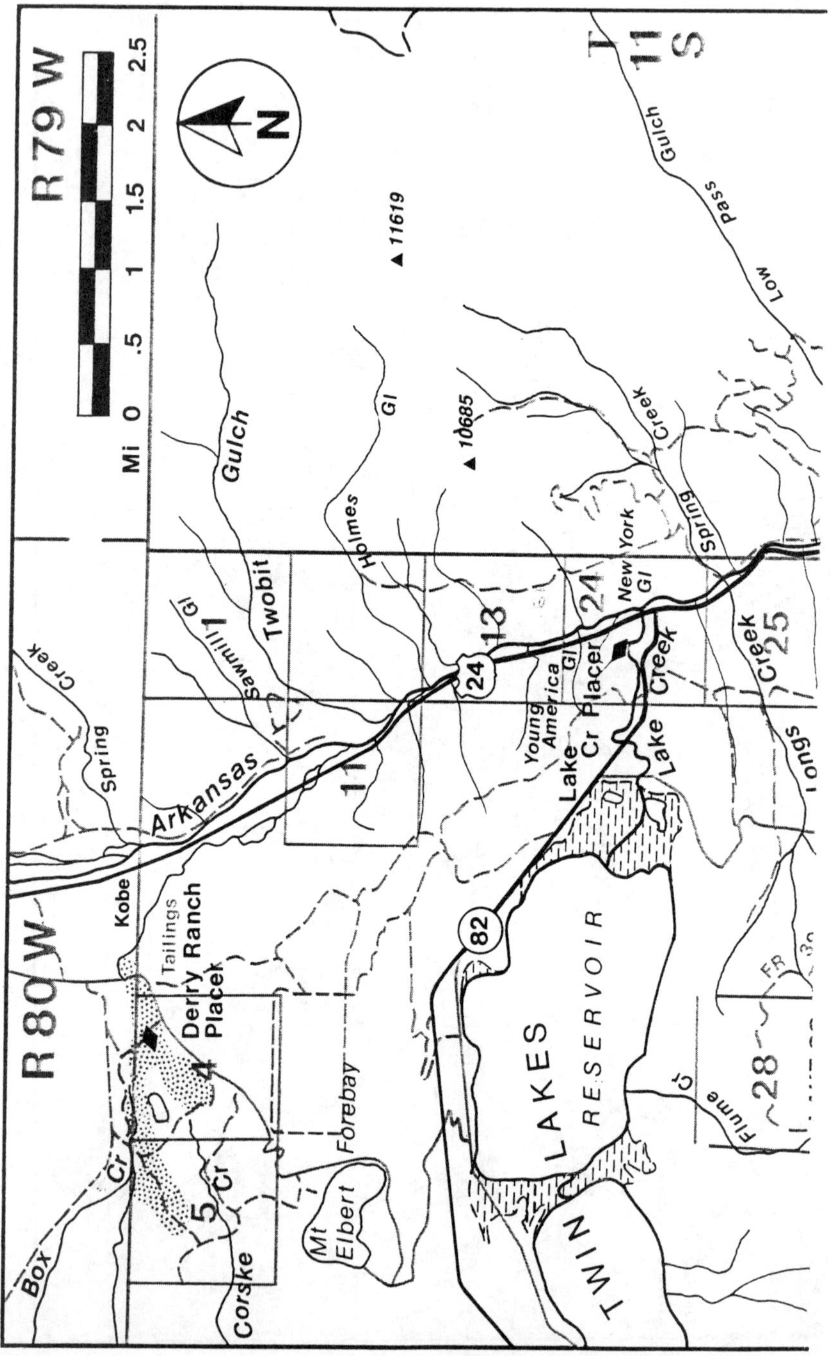
R 80 W
R 79 W
Mi 0 .5 1 1.5 2 2.5
N
T 11 S
Box
Cr
Corske
Cr
Kobe
Tailings
Derry Ranch Placer
4
5
Forebay
Mt Elbert
Spring
Creek
Arkansas
Sawmill Gl
1
Twobit
Gulch
11
13
24
Holmes
Gl
10685
11619
Young America Gl
Lake Cr Placer
New York Gl
Lake Creek
24
25
Spring
Creek
Low Pass Gulch
Longs
Creek
82
TWIN
LAKES
RESERVOIR
Flume Cr
28
FR

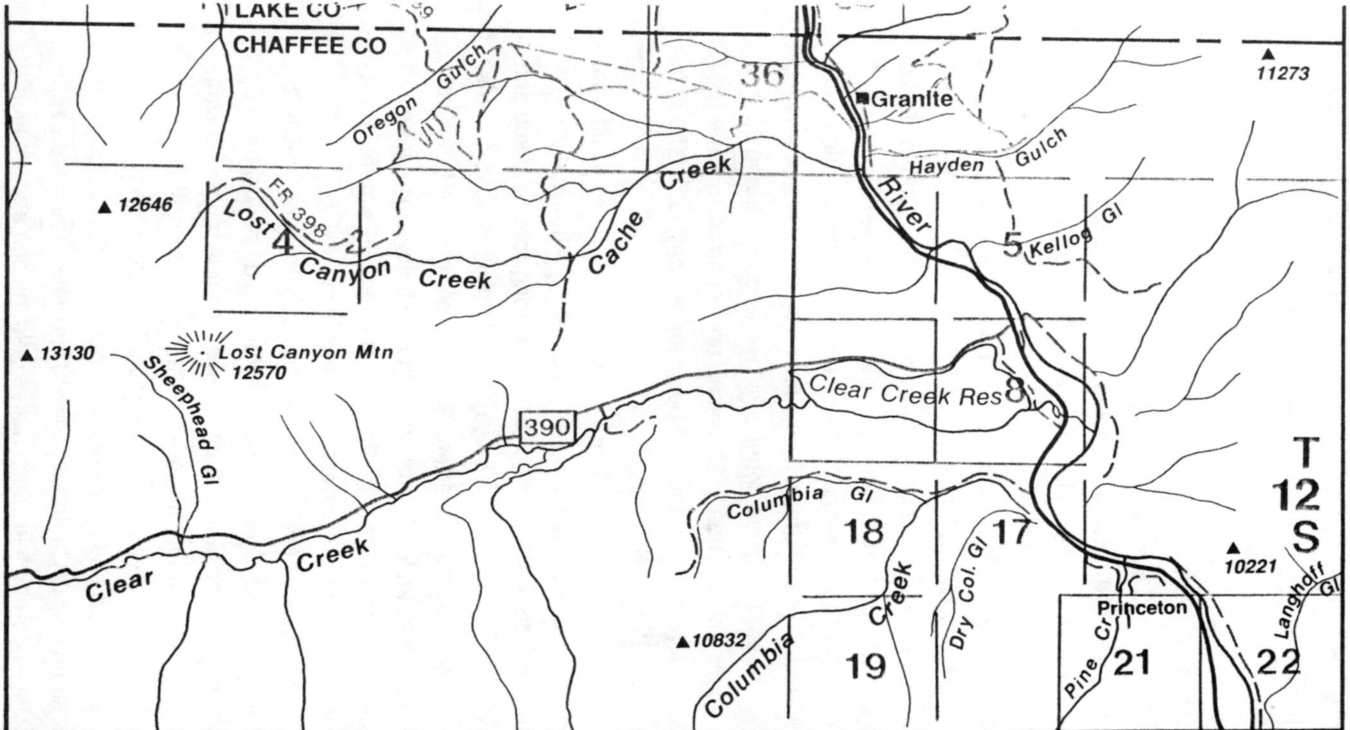

Figure 42. Index map of Derry Ranch and Cache Creek placers, Lake and Chaffee County. (Modified from U.S.G.S. topographic maps)

south of Leadville, where it extends from Malta to Big Union Gulch. Placers are locally associated with some of the terraces.

CALIFORNIA GULCH

Lake County's California gulch and Georgia Gulch in Summit County were the richest and, considering their size, the most productive placers in the State. The total production of California Gulch is estimated to have been $2.5 million to $3 million. At the reported values of its gold. $17 to $19 per ounce, the production was 132,000 to 176,000 ounces.

In late 1859 prospectors found the first indications of gold in California Gulch at its junction, with the Arkansas River, but it was April 1860, before the rich placers in the upper gulch were discovered. One hundred foot claims along the gulch amounted to 339. The gravels in the upper gulch were spotty. During the first eight years, the upstream 92 claims are reported to have paid variously from nothing and $10 per man per day to $100,000 in six weeks (at $17 to $19 gold).

According to tradition, recovery of gold in California Gulch was troublesome because of the heavy "blue sand" that remained with the gold in the pan, rocker, long tom or sluice and that required extra effort to be separated from the gold. Surprisingly, it was not until 1874 that A. B. Wood identified the "blue sand" as cerussite, lead carbonate, and the following year he and Will Stevens discovered the first two of Leadville's great silver-lead orebodies, one on either side of the gulch.

The source of the gold was probably the Printer Boy deposit on Printer Boy Hill. The placer workings do not continue upstream in California Gulch above it, nor are there placer workings in the draws or on the hillslopes on the north side of the gulch. However, there are placer workings in Eureka Gulch (southeast of the site of Oro) and on the hillslope below the Printer Boy mine on the south side of California Gulch. This mine developed a zone of fractured porphyry in which native gold, galena and sphalerite occurred

California Gulch was placered from the Printer Boy downstream to the south side of Leadville, just above the mouth of Georgia Gulch. In the upper gulch the gravels were reported to have been richest at the bend below the site of Oro at the mouth of Nugget Gulch, in the bend about 1,600 feet below Nugget Gulch and in the bend below Graham Gulch near the portal of the Yak Tunnel. From a point about 1,600 feet downstream from the Yak Tunnel the grade of the gravels dropped abruptly. From Georgia Gulch to the Arkansas, although it has been extensively tested, only a little small-scale placering has been done in California Gulch.

From about the Yak Tunnel upstream, the placers were worked to bedrock. Downstream bedrock was too deep to reach and the placers apparently bottomed in compact gravels of Terrace No. 1. In the vicinity of Oro the early miners found a freccia of angular fragments of porphyry cemented with limonite--mineral rust. This lay on bedrock and formed a false bedrock above it. The breccia is gold-bearing. Fragments of this breccia can be found in the boulder piles.

Terrace No. 1 gravels occur on the hills to either side of California Gulch, and the Terrace appears to have extended across and buried the gulch at one time. If this is so, it is possible that the downstream bedrock course of old California Gulch is still covered with terrace gravels; that it leaves the present course of California Gulch near the point that the placer workings no longer reached bedrock and near where the grade of the gravels abruptly dropped. If so it may extend, undiscovered, somewhere to the southwest.

The California Gulch deposit is an alluvial placer. If it was once buried beneath Terrace No. 1, the upper part of the gulch, where bedrock was reached in the placers, is older than the Terrace, or pre-early Bull Lake. The lower part, where terrace grav-

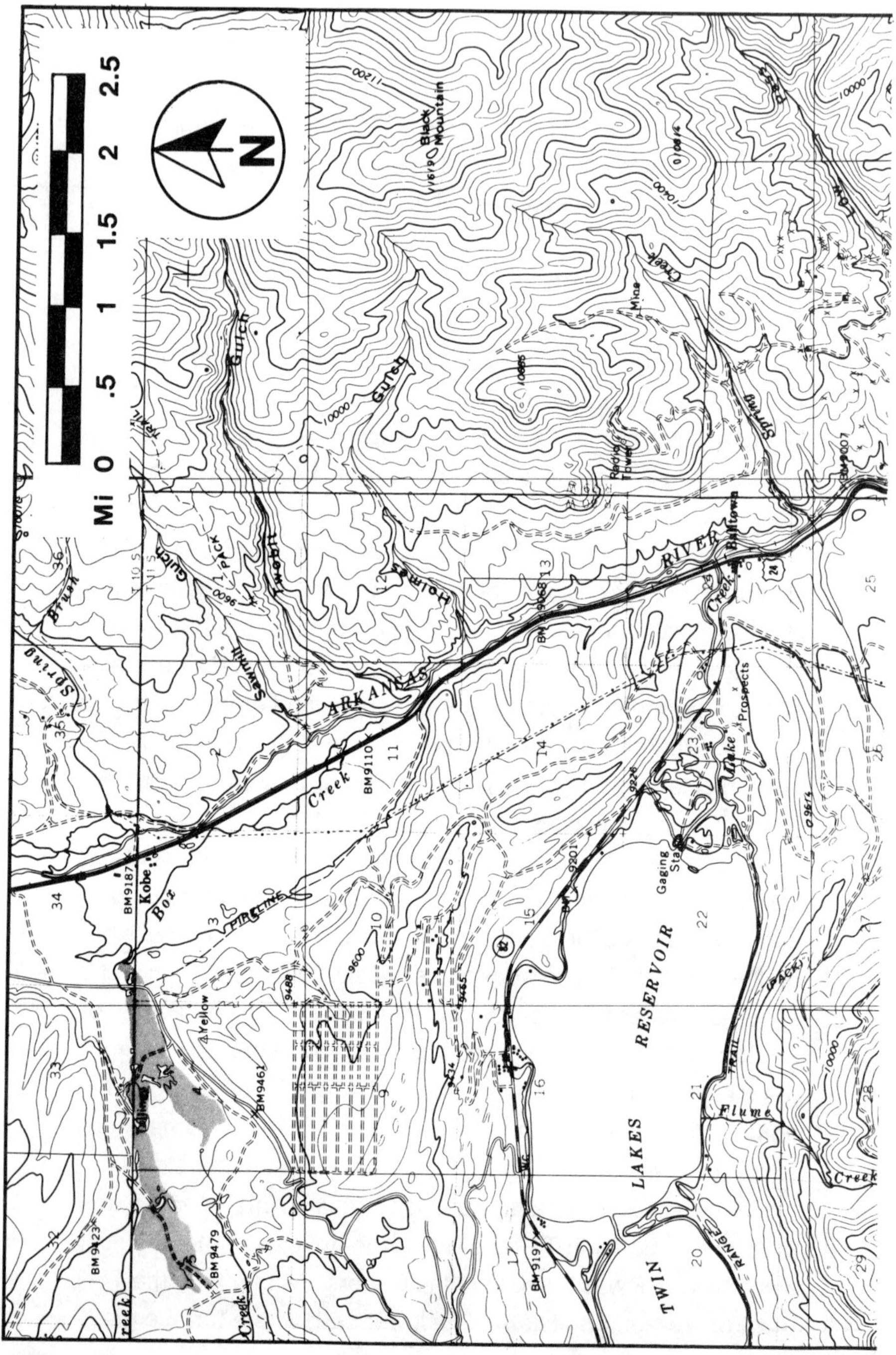

Mi 0 .5 1 1.5 2 2.5
N
Black Mountain
ARKANSAS RIVER
TWIN LAKES RESERVOIR
Gaging Sta
Balltown
Prospects
Flume Creek
Box Creek
Kobe
Yellow
Radio Tower
Mine
Spring Creek
Sawmill Gulch
Twobit Gulch
Holmes Gulch
PIPELINE
RANGE TRAIL (PACK)

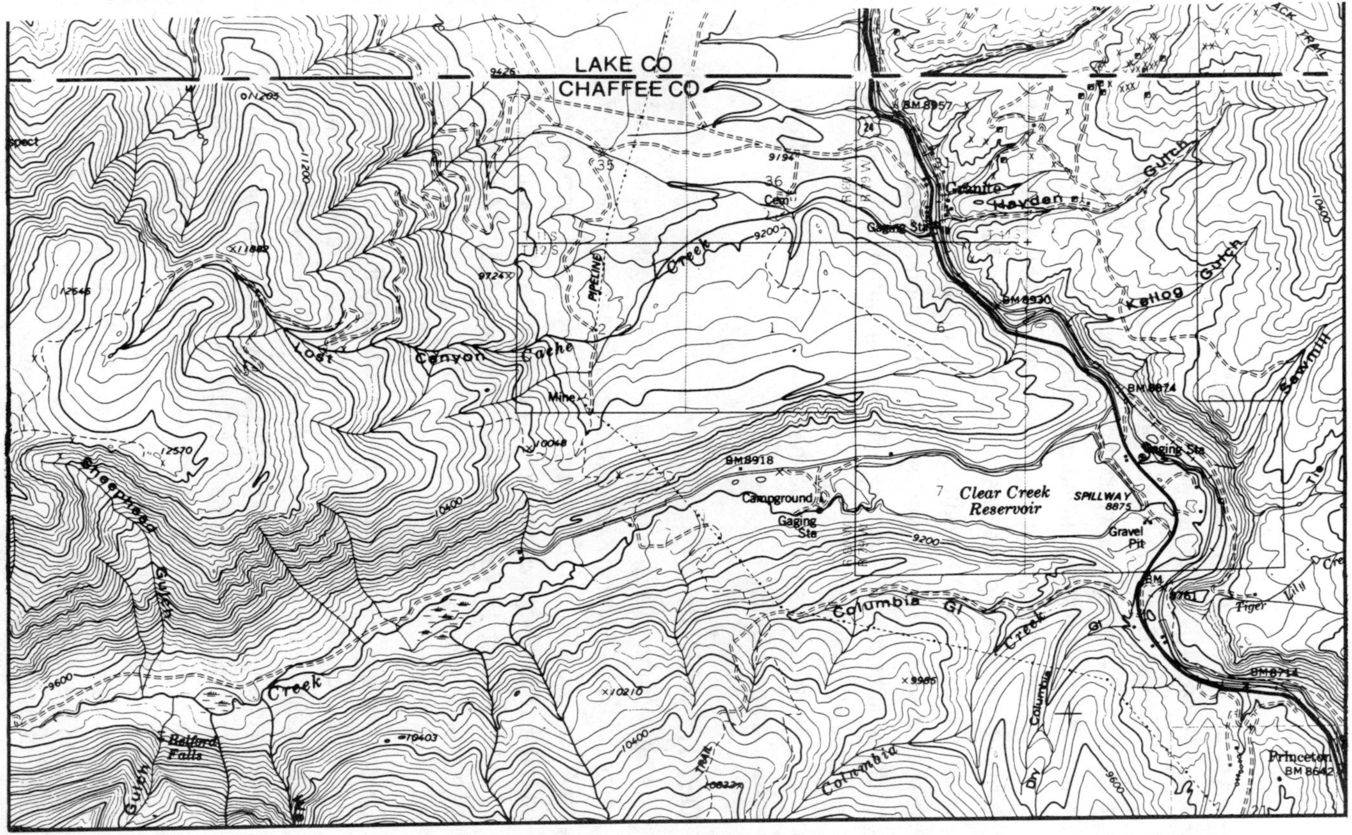

Figure 43. Topographic map of Derry Ranch and Cache Creek area, Lake and Chaffee County. (U.S.G.S. County Map Series)

els formed, the working bedrock is much younger and has formed since Terrace No. 1 was stripped away.

Upper California Gulch has been extensively placered and there is very little unworked gravel, but the old-timers may have left or spilled some on bedrock from place to place.

If you select panning gravel here, select it from the bottom of banks where gravel is exposed or dig down to bedrock beneath the debris. At Oro drift mining was done and a few--very few--pillars may be left. If you encounter the limonite breccia near bedrock, lift it down to bedrock and pan the material on it, with it and beneath it on bedrock.

A few colors can be washed also from deep gravels from lower Califnoia Gulch.

DERRY RANCH

The Derry Ranch placer is on Box and Corske Creeks, to the west of the Arkansas River at Kobe station and some two miles north of the Twin Lakes Reservoir. It appears to have been discovered after the 1860's, and was worked by dredging and by a dry land plant for at least 17 years between 1915 and 1950. More than 68,837 ounces of gold have been produced.

The deposit lies in upper Bull Lake outwash and the modern valley floors incised into it along Box and Corske Creeks, on the northern side of the Lake Creek Bull Lake terminal moraine. The immediate source of the gold is this glacio-fluvial placer is Bull Lake glacial moraine, from Bartlett Gulch and from Lake Creek. The original source is probably the Gordon-Tiger group of veins and others on the slopes of Parry Mountain. The gravels worked in the placered area were up to 50 feet deep. Near-surface colors can be found in pits in SW sec. 4 and SE and W sec. 5, T. 11 S., R 80 W., above the heads of the placered areas.

CAVE CREEK PARK

The Cache Creek placer was discovered in 1859. It is noteworthy for a number of reasons: 1) It was a substantial producer. Most of Chaffee County's production must have come from this placer, so it yielded in excess of $1 million at the old $20.67 per ounce price. No information is available regarding the fineness of the gold. The production had to have been in excess of 49,000 ounces. 2) It was the State's only profitable large hydraulic mine. 3) It was probably the longest operated single placer deposit in the State. From 1859 until 1883 it was worked by individuals and small groups. From 1884 until 1911, when hydraulic mining was stopped by injunction, the mine was operated each season. From 1911 through 1917 placer mining operation decreased, ceasing in 1918. The mine was in operation for nearly 60 years. Subsequently there has been only occasional, small-scale placer mining.

Cache Creek Park is bounded on the north by the Bull Lake terminal moraine of the Lake Creek glacier. The park is bounded on the west by Lost Canyon Mountain and the east by a granite ridge. This ridge rises only slightly above the level of Cache Creek Park to the west, but the Arkansas River has cut a deep canyon below its crest on the east side, within the park, Long's Gulch is on the south edge of the Lake Creek moraine. Oregon Gulch is its southern tributary. Still farther south Gold Run and Cache Creek once flowed across the Park; the placer cut has destroyed their valleys Cache Creek changed its name at the foot of Lost Canyon Mountain; on the Mountain it is call Lost Canyon Creek.

The faces in the placer cut are up to 61 feet high. At many places they have a distinct bench about midway in the gravel exposed. This was a false bedrock. The cobbles and boulders above it are sounder than those beneath it, suggesting that the lower ones have been weathered for a much longer time. The gravel below the bench is lower Bull Lake outwash. That the bench is not present in all the faces shows that the upper Bull Lake outwash streams did not flow over all the Park but were restricted to limited channels

The placer was worked in two "lifts;" the working bedrock of the first lift was about the level of the benches. The bedrock of the second lift, the present floor of the cut, is soft, deeply weathered granite on the east side of the cut and gray and orange clay in the center and west side. In Bull Lake time the glaciers of Lake Creek and Clear Creek pushed across the Arkansas River valley and up its lower east slope. The river was diverted off its valley floor and it cut its present canyon in the granite. From the west edge of this canyon the granite surface slopes downward beneath the moraines toward the mountains on the west. This can be seen in the placer cut, where the granite bedrock sloped downward as the cut is entered from the east. From midway in the cut to its western sides, gray and orange clays, sands and conglomerates, called the "Lake Beds," overlay the granite. The pre-Bull Lake valley floor of the Arkansas River must have been cut in these beds and the lower Bull Lake outwash gravels were deposited on them. (They form the bed-rock of the Derry Rancy placer to the north of Twin Lakes.)

Cache Creek Park had two sources of gold: The principal source lay to the west or southwest--probably Lost Canyon Creek--and the moraines of Lake Creek also provided some gold.

Cache Creek Park was an outwash terrace of lower Bull Lake age, 155 feet above the Arkansas; its gravels contained gold from both sources. In upper Bull Lake time these gravels were reworked by streams which brought in more gold from both sources and additional gravel. Gold Run (destroyed by extensive placer workings) and Cache Creek must have been sites of such reworking and deposition of upper Bull Lake outwash; their valleys were destroyed by the placer mining. Long's Gulch and Ritchey's Patch were the sites of similar reworking and Long's Gulch probably did not receive gold from Lost Canyon as did the other areas and so the placer deposits in it are much less important. Gold Run and CacheCreek continued to receive gold from Lost Canyon after Bull Lake time as Pinedale and more modern streams reworked some of the upper Bull lake outwash.

GUNNISON COUNTY

North Taylor Park and Bertha Gulch have been only small placer gold producers but they are of special interest because their placers carry some heavy minerals besides gold and magnetite. They are in northeastern Gunnison County in the Upper Taylor River area.

The first reports of mining in North Taylor Park, at Kent Gulch, were in 1867; probably work began in the nearby gulches about the same time. The first report of placering in Bertha Gulch was in 1881. Desultory mining continued at both places through the 1880's. In 1896 the Taylor Park Placer Co. completed its ditch, bringing water from the Taylor River to Little Taylor Gulch, and placering continued in North Taylor Park for a few years afterward. There was a little activity again in Depression times.

Most of the work in these two areas was manual--rockers, booming and shoveling-in augmented by small-scale hydraulicking where water was available. The Placer Company began construction of a bedrock flume at the Taylor River to work Illinois Gulch. The work was abandoned in 1899 when the bedrock gradient was found to be too low to operate a flume. (This was also the final result of other bedrock flume projects in the district at Little Taylor Gulch and Union Canyon.) The only other large scale mining attempted was the operation of a hydaulic elevator in Illinois Gulch in 1898.

NORTH TAYLOR PARK

These placers lie in Illinois and Pieplant Creek and their tributaries' valleys below 9850 feet elevation on Pieplant Creek and below 9750 feet on Little Taylor Gulch. Taylor River and Illinois Creek have broad, outwash filled valleys. From place to place along Illinois Creek there are terraces 85 to 95 feet above stream level.

Placering was attempted in the outwash plain at the mouths of Illinois Creek and of Little Taylor Creek, but the workings appear not to have reached bedrock. Placers have been worked from place to place in most of the tributary gulches between Illinois and Pieplant Creeks, in a few spots on the terraces and on the hillslopes and in ravines on them. The gulch placers and those on the slopes and ravines appear to have been profitable; the terraces were little worked and apparently offered encouragement to the miners in only a few places. In the gulch placers the gold bearing gravels were covered with barren debris which thickened upstream. The gulches were worked to where it became uneconomical to move this cover.

On the hill slopes and in the ravines the gravels were usually covered, but with no more than three feet of saprolite--granular, disintegrated granitic rock debris.

The gold found in these placers is reported to be small flakes, wires and, rarely, small nuggets. With the black sand in pan concentrates there frequently is minor columbite-tantalite, monazite, zircon and garnet. Columbite-tantalite is an ore of tantalum; it appears as black sand, resembling magnetite but is nonmagnetic. Monazite is a thorium-bearing mineral; it appears as blackish to greenish grains and is slightly radioactive. Zircon appears as yellow to orange grains and garnet most commonly as red to orange grains. These minerals are not abundant in the concentrates; they have never been produced commercially.

The source of these other heavy minerals is the granites and the pegmatites in the area, but the source of the gold has never been identified. The placers are widespread; this suggests that there are a number of sources. The placers are below the Taylor Park Placer ditch. Whether were no gold-bearing gravels above it or that none were worked above it because of lack of water is unknown.

Gravels for panning can be found in some old workings in the gulches and ravines. Debris on bedrock in old hillslope placer areas should yield some colors and any gravel found on the slopes should be panned. The terraces are probably low grade. secs. 27 or 28 should be good places to start.

BERTHA GULCH

There is an extensive hydraulic cut in the fan gravels at the mouth of Bertha Gulch and small cuts from place to place upstream across sec. 1,T. 15 S., R. 82 W. The gravels of the hydraulic cut reportedly were better than those upstream; the best of the upstream gravels were just west of the center of sec. 1.

Scheelite, an ore of tungsten, occurs in the heavy minerals of the upstream gravels. Its recognition in pan concentrates led to the discovery of gold-scheelite-quartz veins on the north slopes of the gulch in sec. 2. These veins did not prove to be economic, but they are a bedrock source for the gravel here. Some placer gravels are still exposed in the hydraulic cut and a few places upstream.

Similar bur much smaller placers were worked in Kentucky Gulch, which lies 1 to 1 1/2 miles to the north. These placers are ones that you may have to prospect some to come up with a good pan of gravel.

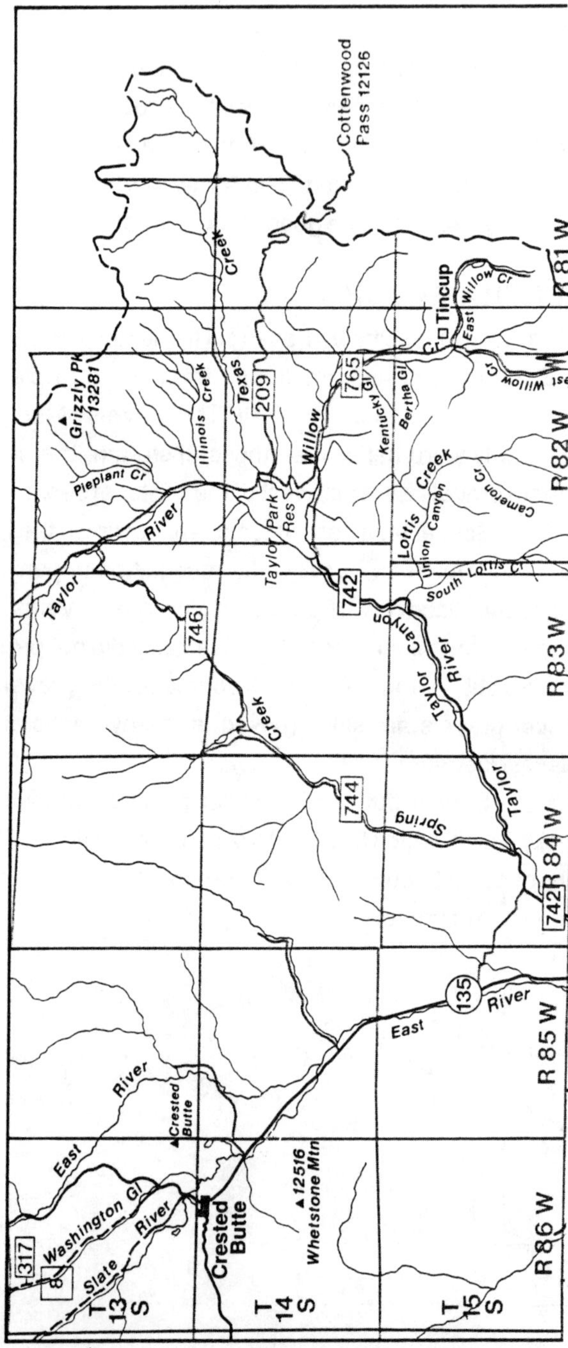
Cottenwood Pass 12126
Grizzly Pk 13281
Pieplant Cr
Illinois Creek
Texas Creek
209
Willow
765
Kentucky Gl
Bertha Gl
Cr
Tincup
East Willow Cr
West Willow Cr
R 81 W
R 82 W
Lottis Creek
Union Canyon
Cameron Cr
South Lottis Cr
Taylor Park Res
Taylor River
742
746
Taylor Canyon
Taylor River
R 83 W
Spring Creek
744
R 84 W
742
135
East River
R 85 W
Crested Butte
East River
12516 Whetstone Mtn
Crested Butte
Washington Gl
Slate River
317
8
T 13 S
T 14 S
T 15 S
R 86 W

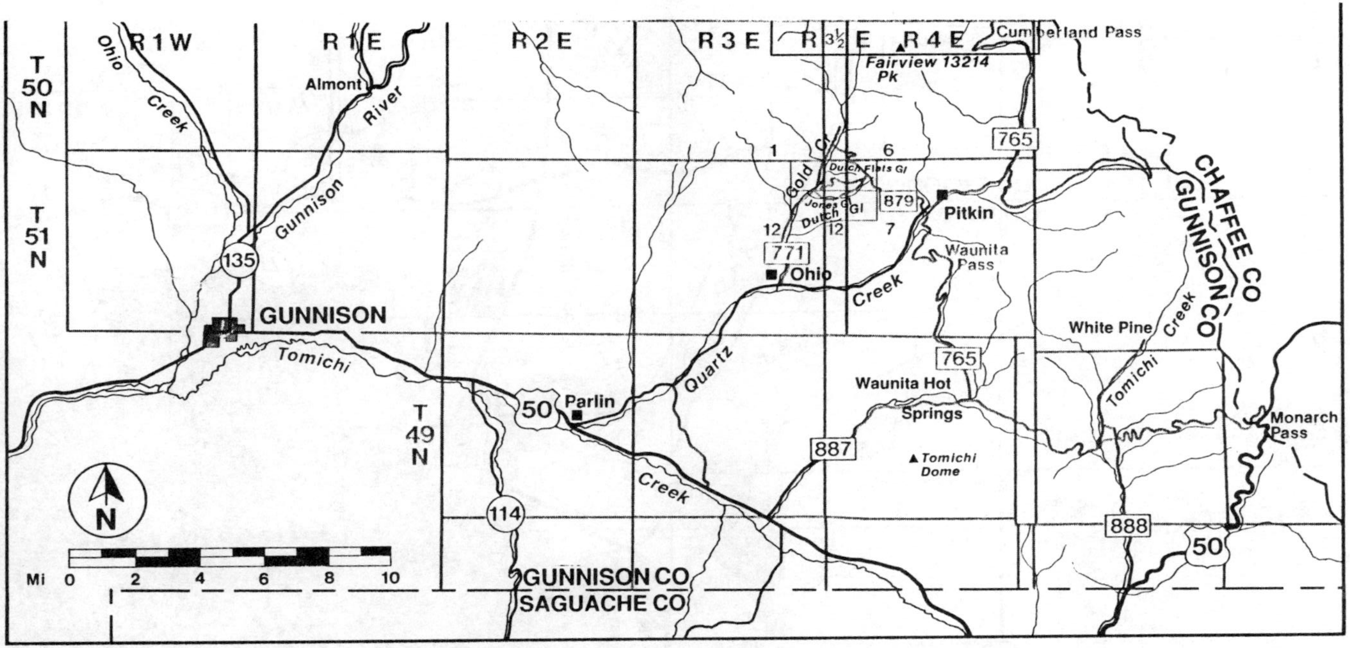

Figure 44. Index map of Gunnison County placers. (Modified from U.S.G.S. topographic maps)

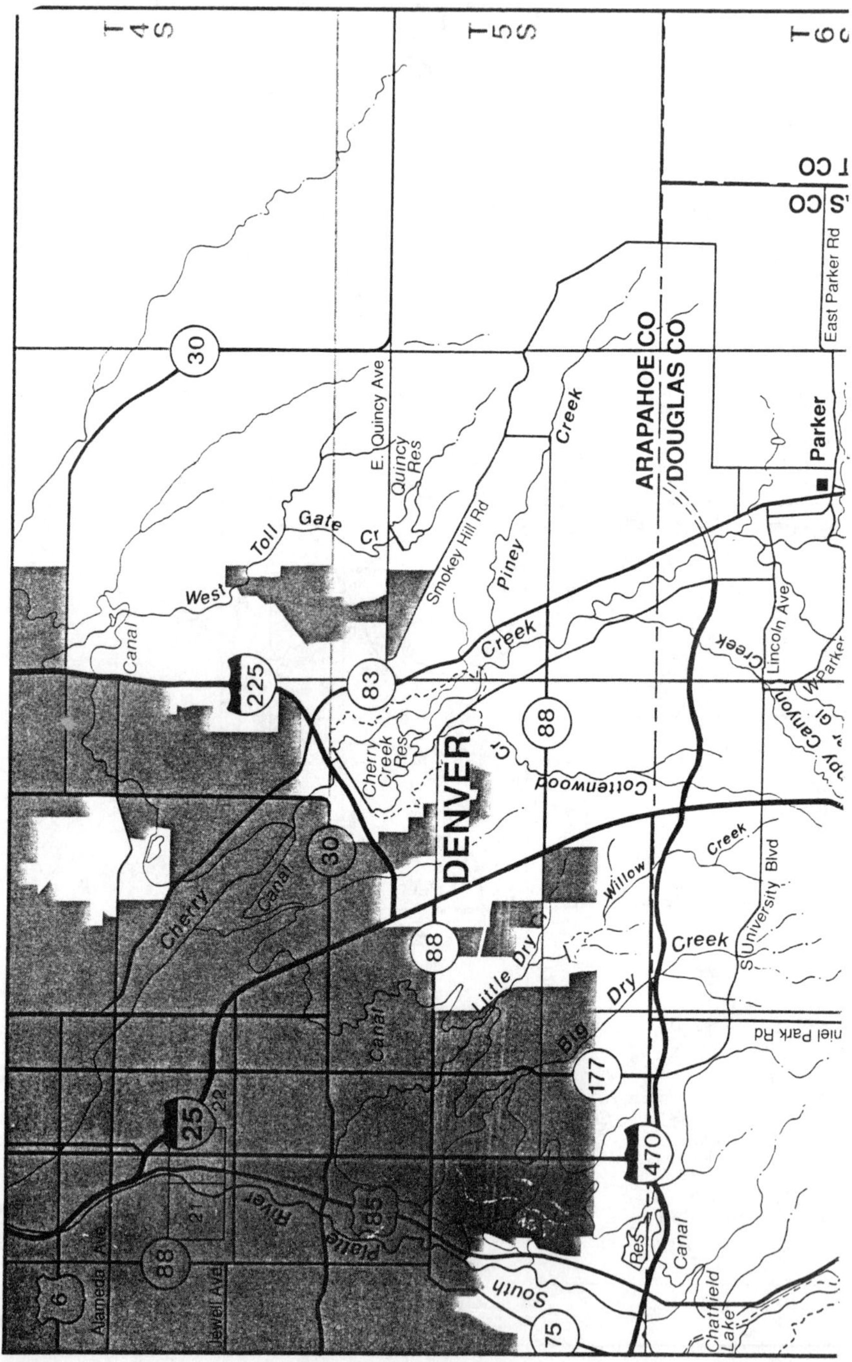

T 4 S
T 5 S
T 6 S
CO
CO
East Parker Rd
30
E. Quincy Ave
Quincy Res
West Toll Gate Cr
Smokey Hill Rd
Piney Creek
ARAPAHOE CO
DOUGLAS CO
Parker
Canal
225
83
Creek
Lincoln Ave
Cherry Creek Res
DENVER
Cr
Cottonwood
88
30
Cherry
Canal
Willow Creek
S University Blvd
88
Little Dry Creek
Big Dry Creek
Canal
177
25
22
470
21
Platte River
85
88
Alameda Ave
Jewell Ave
6
South
Res
Canal
75
Chatfield Lake

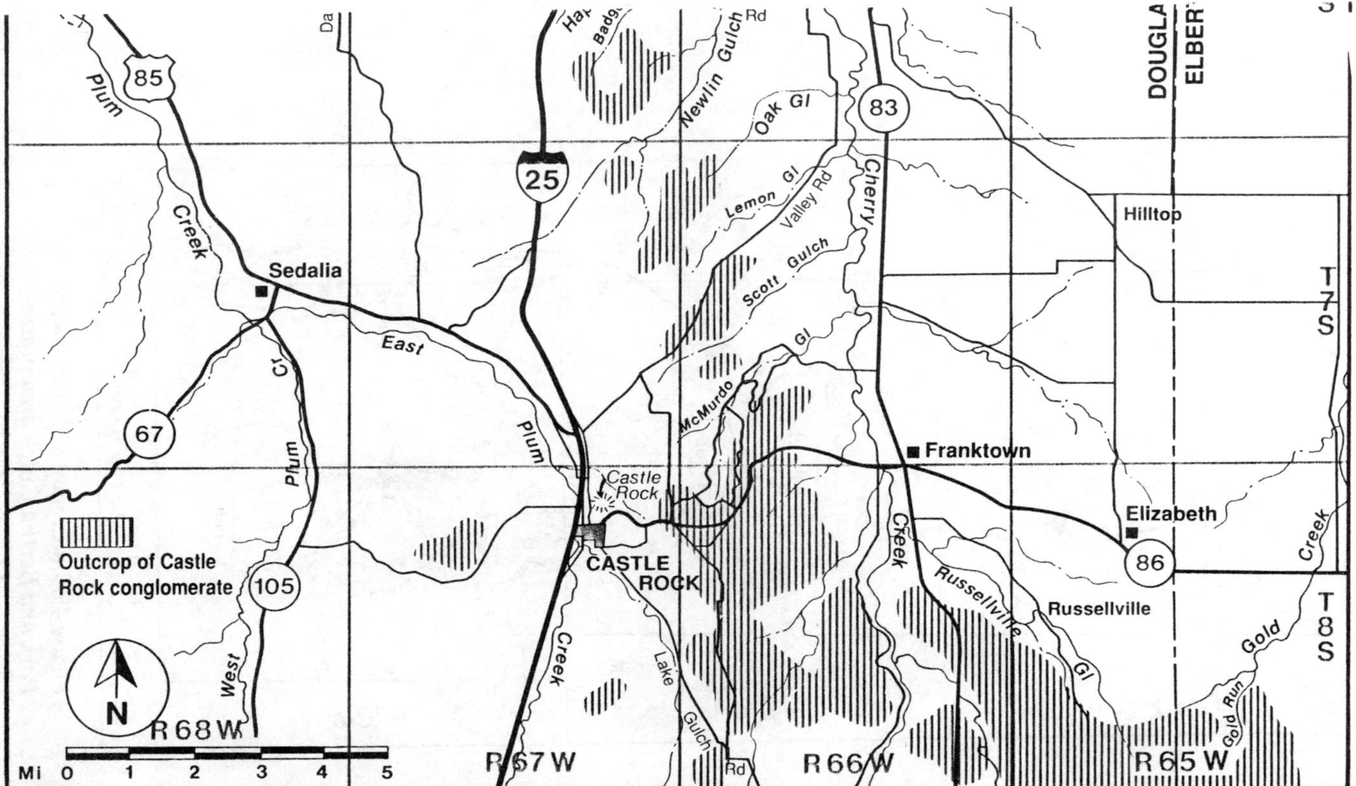

Figure 46. Index map of the Cherry Creek Divide area placers, Arapahoe, Denver, Douglas, and Elbert Counties. (Modified from U.S.G.S. topographic maps)

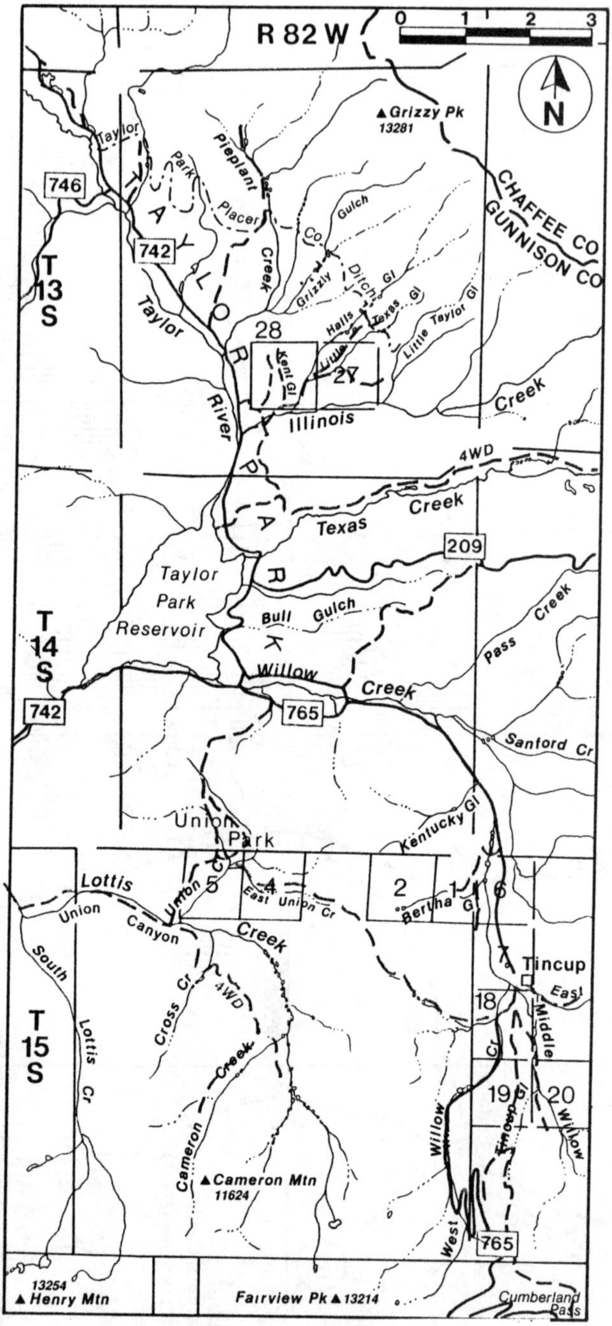

Figure 45. Index map of North Taylor Park and Bertha Gulch area placers, Gunnison County. (Modified from U.S.G.S. topographic maps)

CHALLENGING AREAS FOR PANNING

CHERRY CREEK DIVIDE AREA

ARAPAHOE, DENVER, DOUGLAS, AND ELBERT COUNTIES

This area includes the sites of the gold discoveries of 1858, which led to the "Pikes Peak or Bust" gold rush of the following year. The first worthwhile deposit was found on the South Platte River, in Denver, County, probably in sec. 21 and 22. T. 4 S., R. 68 W., (near Overland Park) between the present 8th and Jewell Avenues. The most productive deposits found later were on Big Dry, Newlin and Russellville Gulches in the Cherry Creek drainage in Douglas County and on Ron (now called gold Creek) and Gold Run Gulches to the east of Russellville Gulch in Elbert County.

The placers were pretty much abandoned by the end of 1859; the miners left for the richer placer areas discovered elsewhere that year. However, there has been both gulch and drift mining from time to time since, especially individual and small-scale operations during Depression times.

The gulches in which the placers lie drain the Cherry Creek Divide, which is capped by the Oligoncene Castle Rock Conglomerate. (This formation forms "Castle Rock," overlooking the town of the same name.) To the northeast of Castle Rock town the conglomerate has been eroded by the streams and now is represented only by remnants capping buttes along the divide. Many hillslopes have residual pebbles and cobbles on them, derived from the erosion of the conglomerate. There are terraces or benches at many places along the streams. Along the South Platte the terraces are 40 feet above the level of the river. These benches extend into Cherry Creek and its tributaries, their height diminishing upstream.

Gold has been found in the terraces and in the stream beds. Some testing has shown that, at least locally, gold in the bench placers lies on or near bedrock but along the stream beds it is nearer surface. In most of the productive gulches gold values are spotty. Along Happy Canyon and Big Dry Gulch reportedly gold is widespread but values are very low. In some gulches, among them Badger, Oak and Cottonwood Gulches and the Plum Creek drainage, no gold has been found.

The Castle Rock conglomerate contains a little gold at many places. In upper Newlin Gulch near the forks some small, lens shaped beds of gravel were found in it that reportedly contained up to one or more ounce of gold per ton in individual samples. After their discovery in 1895, these were worked in several small drift mines. The lenses were 5 to 40 feet above modern stream levels, probably near the base of the Castle Rock formation.

The gold in all the placers was typically fine-grained--10 to 50 colors per milligram. Miners described it as "almost flour gold" and complained of the difficulty of its recovery from the clayey gravel matrix. Even so, colors up to one-sixteenth inch diameter were occasionally found. The gold is unusually pure, characteristically about 990 fine,

The immediate source of the stream and terrace placer gold is the Castle Rock conglomerate and fossil stream placers in it. That the source is fossil stream channels does much to explain the erratic distribution of gold among the gulches and in each of them.

Access to the placer gravels is difficult because of the many developments in the area. There is very little water and the gravels must be packed out. Remember to get permission to enter and to leave gates as you found them, open or closed!

The gravels of Cherry Creek and of the South Platte below the mouth of Big Dry Gulch contain some gold but are very low grade and spotty. Some samples of the Castle Rock conglomerate, particularly of coarse, cobbley sections that may be old stream channels, will show fine colors. Perhaps the best stream areas are Newlin Gulch at and below its forks and Russellville, Ronk (Gold

Creek) and Gold Run Gulches and other gulches in T. 8 S., R. 65 W., near and below the base of the Castle Rock Conglomerate.

SOUTH BOULDER CREEK AND TRIBUTARIES

BOULDER AND GILPIN COUNTIES

From the junction of Beaver Creek with South Boulder Creek for about 3.5 miles downstream South Boulder Creek is the boundary between Boulder and Gilpin Counties. The Deadwood Diggings, the third of the economic gold discoveries made early in 1859 that maintained the Colorado gold rush, was on South Boulder Creek near the mouth of Beaver Creek. Gold-bearing gravels probably extended downstream for some distance, but little evidence remains of early day activity there.

GILPIN COUNTY

Within Gilpin County the gravels of South Boulder Creek are gold-bearing from the mouth of Moon Gulch to Beaver Creek. On South Boulder Creek there are two parks where the valley floor widens, one centered on Rollinsville and the other centered on Pactolus Station and extending to the Deadwood Diggings area. Both were worked with dry land washing plants in the 1930's. Earlier placers were worked in Lump Gulch from near where Colorado 119 approaches it from the south upstream for about 3/4 of a mile and in Gamble Gulch from near where Colorado 119 crosses it upstream to the site of the old village of Perigo. These deposits are stream placers. Some gravel still can be found in their banks from place to place.

UPPER CLEAR CREEK

GILPIN COUNTY

Small placers were worked along North Clear Creek and Missouri

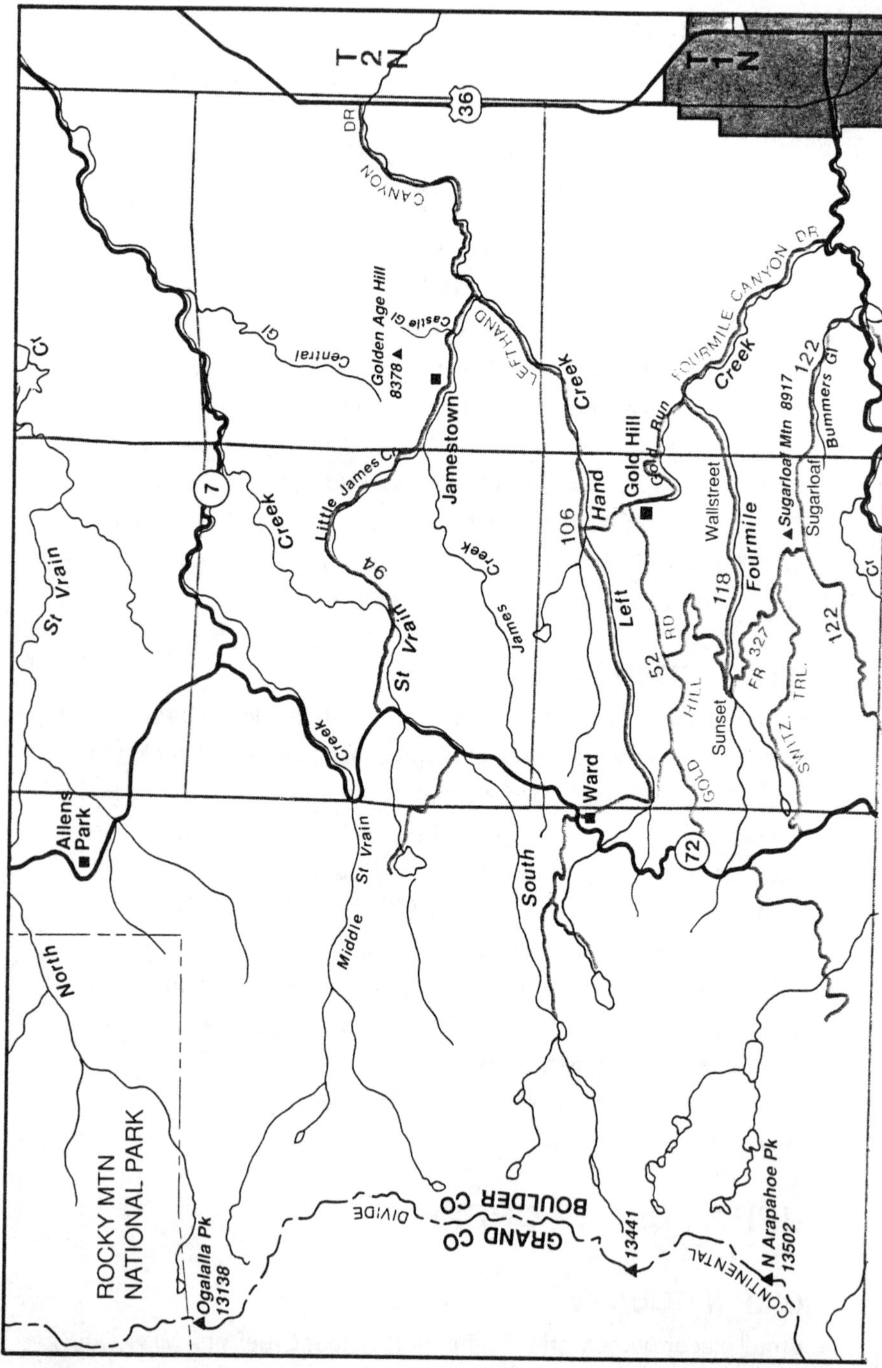
ROCKY MTN
NATIONAL PARK
Ogalalla Pk
13138
North
Allens
Park
Middle
St Vrain
South
St
Vrain
Creek
7
94
St Vrain
Creek
Little James Cr
Jamestown
James
Creek
Golden Age Hill
8378
Central
Gl
Castle Gl
LEFTHAND
CANYON
DR
36
T
2
N
T
1
N
Ward
Left
Hand
Creek
106
Gold Hill
Gold Run
52
GOLD
HILL
RD
Sunset
Wallstreet
118
Fourmile
Creek
FOURMILE CANYON DR
72
FR 327
SWITZ.
TRL.
Sugarloaf Mtn 8917
Sugarloaf
Bummers Gl
122
122
Cr
Cr
BOULDER CO
GRAND CO
DIVIDE
CONTINENTAL
13441
N Arapahoe Pk
13502

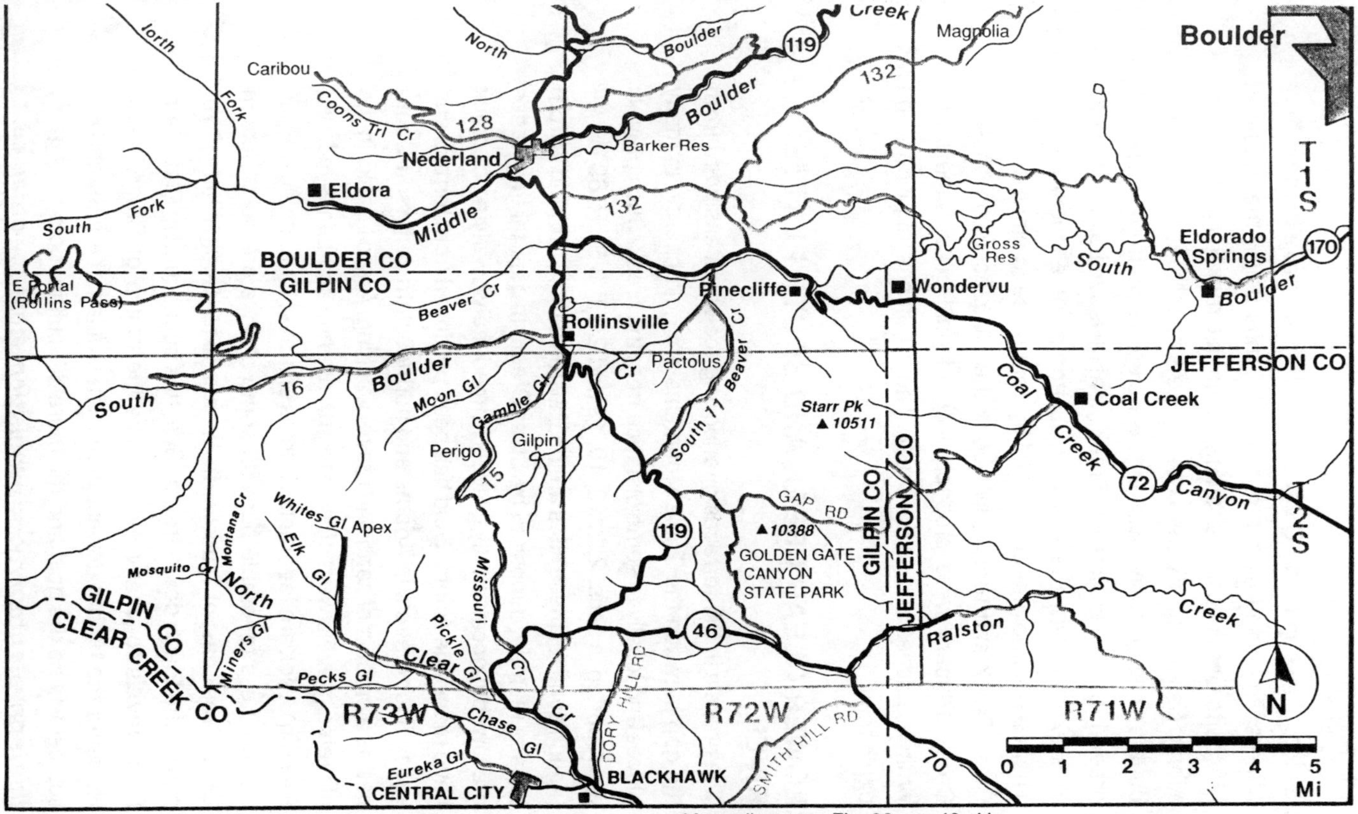

Figure 47. Index map of placers in South Boulder Creek and tributaries area, Boulder and Gilpin Counties; and the Upper North Clear Creek placers, Gilpin County. (Modified from U.S.G.S. topographic maps)

Gulch. Those in North Clear Creek are just above the mouth of Miners Gulch and from a quarter of a mile below the mouth of Mosquito Creek upstream along Montana Creek to near its head. Along Missouri Creek there were placers near North Clear Creek, but the most extensive workings were upstream from where Colorado 119 crosses Missouri Creek, near Missouri Lake. The workings extend from the highway upstream for about 0.7 miles. There are some gravels remaining from place to place on North Clear Creek and reportedly above the old placers on Missouri Creek. Placers were probably worked on Pecks, Miners and Whites Gulches and on Elk Creek. These gulches might be worth exploring.

LOWER CLEAR CREEK AND SOUTH PLATTE RIVER

The gravels in the terraces overlooking the creek eastward from the gap between the Table Mountains are gold bearing in many places and locally to some distance from the creek. Such gravels are reported in secs. 2, 3, 12, 10, 19 and 24, T. 3 S., R 69 W. Although some of these sites are on Ralston Creek, the gold is from terrace material deposited by Clear Creek. In general the grades are low, but occasionally gravels immediately overlying bedrock have many fine colors. Gold has been produced from terrace and valley floor gravels in Adams and northeastern Jefferson Counties as a product of sand and gravel washing. Any pit banks in the elevated gravels west of Wadsworth avenue and banks in lower Ralston Creek are favorable sources of gravel for panning.

Less favorable but with some chance are point bars in Clear Creek east of Wadsworth Avenue to the South Platte River and on the Platte downstream. Gold has been found in Platte surface gravels for many miles below the mouth of Clear Creek. However the farther downstream from Golden, the finer and the scarcer will be the gold and the more difficult to recover. ***Again, this is private property, so permission should be obtained.***

BOULDER COUNTY

Placer gold has been produced at the "deadwood Diggings" on South Boulder Creek at the Jefferson Diggings" on North Beaver Creek about four miles above South Boulder Creek, along Gold Run south of Gold Hill, along Fourmile Creek from place to place from Boulder Creek to near Sunset, "near Sugarloaf" (probably Bummer's Gulch), on Coon Trail Creek between Nederland and Caribou, on Castle Creek just below its mouth on James Creek, and at the head of Central Gulch northeast of Jamestown. (The last two gulches drain the slopes of Golden Age Hill, below the vein of the same name.) Placer gold is reported in the gravels of Boulder Creek from the west side of Boulder to the mouth of Fourmile Creek and on the North St. Vrain north of Allens Park. The gravels at both these last two sites are probably extremely low grade.

CHAFFEE COUNTY

LOST CANYON

The Lost Canyon placers lie on the northeast slope of Summit 12,570 in secs. 4 and 5 (unsurveyed), T. 12 S., R. 80 W., two miles west of Cache Creek Park. The placers lie in a basin between 11,250 fee and 12,100 feet in elevation. The placers extend from the forks of Lost Canyon Creek upstream along the northern, main branch more than a mile and downstream a short distance.

In 1860 prospectors returning from Taylor Park discovered this deposit. Reportedly they were lost when they found it; hence the name. Summit 12,570 lies at the eastern end of a high ridge 10 miles from the Continental Divide, which forms the eastern limit of Taylor Park. That travelers coming east from Taylor Park would pass by this mountain suggests that they were truly lost or, more likely, that in the then trackless country it was easier to travel above timberline than in the forested valleys.

Operators have told me that although gold is concentrated in a pay gravel on bedrock, this gravel varies from a few inches to 15 feet thick. The pay gravel contains angular rock fragments, with fewer large fragments than the mantle over it, and the largest fragments in it are less than 18 in diameter. It contains little clay and is easy to wash. The gold is coarse and rough; gold is frequently attached to quarts fragments.

This deposit is unusual in several ways: 1) Despite the high elevation of its head, Lost Canyon was not glaciated in Illinoian and Wisconsin time, so the placer is old and has developed since before Illinoian time. 1) Below the placers the valley changes from an open basin to a very steep, narrow valley incised I00 to 150 feet into the mountain slope, and just above the level of Cache Creek Park it changes again to an open, gently sloping valley. These lower sections of the valley appear to have been well prospected but never worked. Reportedly, the pay disappeared a short distance below the forks in sec. 4. It seems reasonable that upper Lost Canyon was a source of some of the gold in southwestern Cache Creek Park, but no placer gold has been found between them. 3) The source of the gold in Lost Canyon has never been recognized. It must lie in the upper basin, but whether one or several good bedrock deposits or a widespread, low-grade occurrence is unknown.

Select material for panning like the pay gravel described above. If there is none, take the material as deep as possible in the cuts. There is some suggestion that mantle away from the stream may contain gold from place to place. Should you wish to try this "long shot," prospect the upper and northern parts of the basin where mantle is thinner. By midsummer water is scarce here and you must "either pack water up or gravel down."

COLUMBIA GULCH

Placers have been worked on Columbia Gulch and Columbia Creek in secs. 17, 18, 19 and 20, T. 50 N., R, T E.

MT SHAVANO

There are placer working on Placer Creek in SWSW sec. 14 and E SE sec. 15, T. 50 N., R 7 E. and at the junction of Placer Creek with Blank Gulch in sec. 26.

The placers are in valleys incised in a piedmont alluvial plain (a plain formed at the foot of mountains by the coalescing of gently sloping alluvial fams and stream deposition between them) at the foot of Mount Shavano. The thickness of the gravel beneath the plain is unknown, but across the South Arkansas valley the gravel beneath a similar plain is more than 200 feet thick. The :bedrock" at the placer deposit is probably undisturbed piedmont gravel. This plain is pre-Illinoian in age. Blank Gulch and Placer Creek began to be incised in it sometime after its deposition and the placer deposits have developed since that time.

The gold produced at these placers is said to have been heavy and rough and the placer gravel to contain a great deal of black sand. The original source of the gold apparently is a mineralized area high on the southeast face of Mount Shavano, where reportedly some eluvial placers were worked years ago.

Select gravels for panning at either end of the workings on Placer Creek. If you like the area, the placer channels (if any) between the Placer Creek workings and the bedrock deposits on the peak have never been found.

Placers are reported on upper Browns Creek, five or six miles to the north in T. 51 N., R. 7 E. If this is true, some deposits yet to be discovered may exist at the foot of the mountain and the top of the alluvial plain in the gulches between Blank Gulch and Browns Creek.

TROUT CREEK

Placers were worked on Trout Creek "one mile from the Arkansas River," probably in secs. 26 and 34, T. 14 S., r. 78 W.

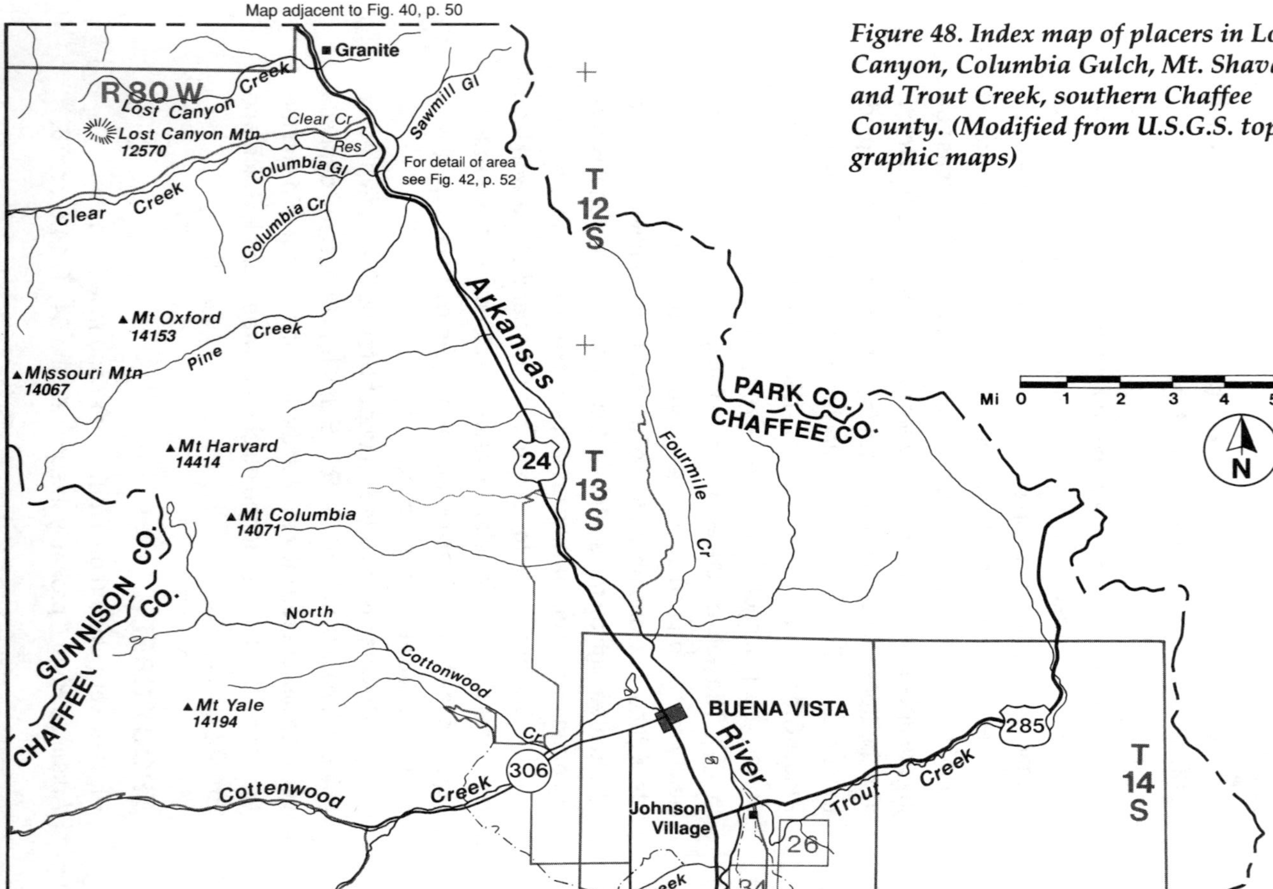

Figure 48. Index map of placers in Lost Canyon, Columbia Gulch, Mt. Shavano, and Trout Creek, southern Chaffee County. (Modified from U.S.G.S. topographic maps)

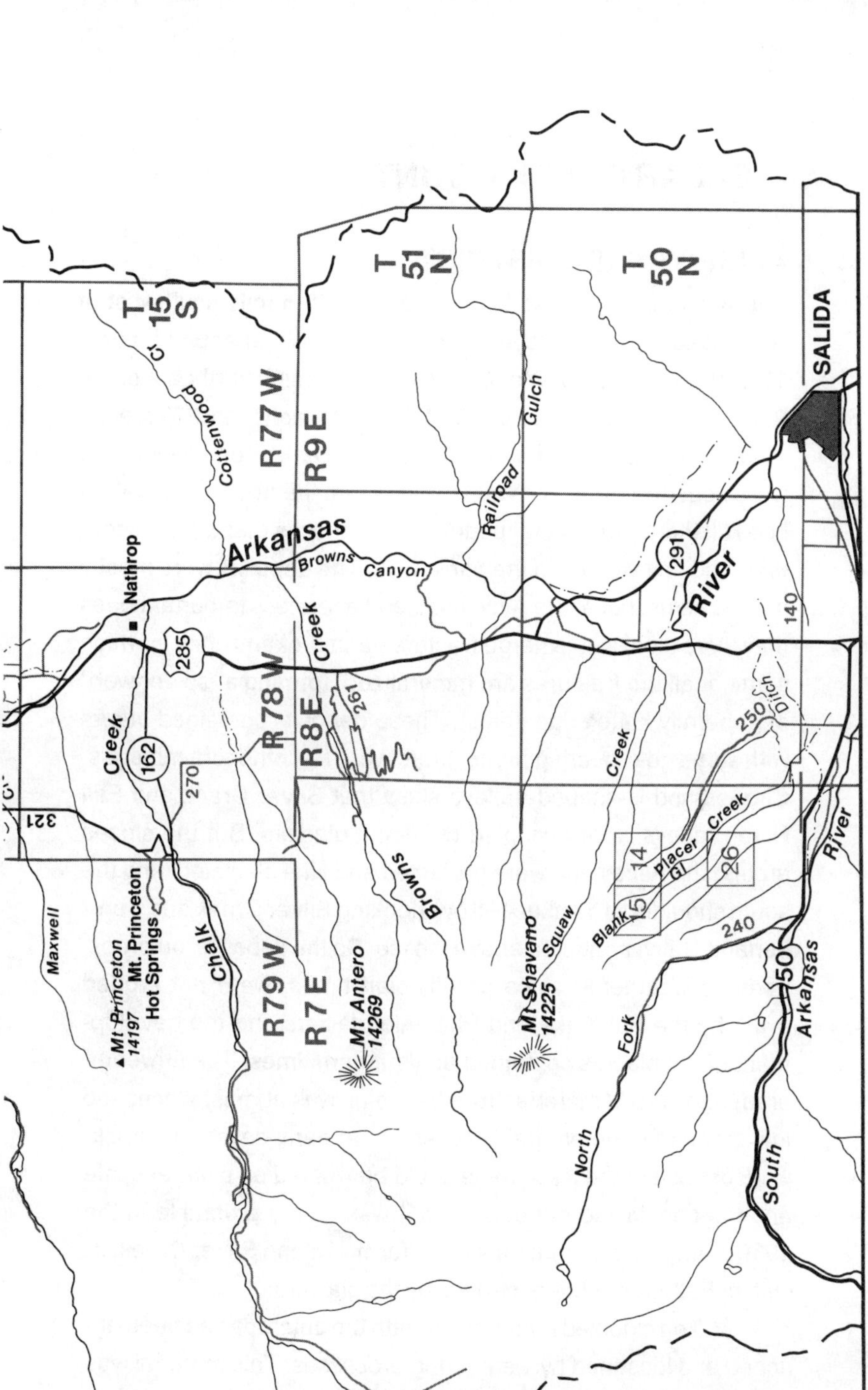
SALIDA
T 15 S
T 51 N
T 50 N
R 77 W
R 9 E
R 78 W
R 8 E
R 79 W
R 7 E
Arkansas
River
Arkansas
River
South
Cottonwood
Cr
Browns
Canyon
Creek
Railroad
Gulch
Nathrop
285
291
162
321
270
261
250
240
140
50
Creek
Mt Princeton
14197
Mt. Princeton
Hot Springs
Maxwell
Chalk
Mt Antero
14269
Browns
Mt Shavano
14225
Squaw
Creek
Blank
Placer
Gl
Creek
Ditch
North
Fork
14
15
26

CLEAR CREEK COUNTY

ALICE/YANKEE DISTRICT

The Alice mine, marked by a large cut, is 0.6 mile southwest of Alice in sec. 3 (unsurveyed), T. 35., R. 74 W., at approximately 10,300 feet elevation. Yankee Hill lies 1.8 miles northeast of the Alice Mine in sec. 36, T. 2 S., R 74 W., and reaches 11,239 feet elevation. The sites of the ghost towns of Yankee and Ninetyfour were near the summit of Yankee Hill. From the Alice mine to Yankee Hill there are several bedrock ore deposits that were discovered and first worked in the early days. The deposits were mostly narrow veins, but at the Alice mine and a few less important ones they were stockworks (large fractured and broken zones in which some or all the fractures are mineralized, forming a "spider web" of generally knife edge veins). These deposits contained quarts with some gold-bearing purite, there were few other ore minerals. Cirques and U-shaped valleys show that Silver Creek and Fall River valleys were occupied by alpine glaciers. But the slopes around the Alice mine were protected and little glaciated, and the south shoulder of Yankee Hill overlooking Silver Creek and Cumberland Gulch projected above the ice. So the deposits' outcrops, perhaps weathered since pre-Illinoian times, were not eroded away by the Bull Lake and Pinedale glaciers and the development of the placers continued until modern times. The unweathered, unoxidized material from the veins was at most places too low grade to mine profitable, even in the early days. The stockwork deposit of the Alice mine could be worked on a larger scale and the unoxidized material from it was mined profitable in the 1930's. Any placers that may have formed along Silver Creek, itself, or Fall River were destroyed by the glaciers.

The oxidized material beneath the outcrops had been enriched and loosened by weathering processes. This material was readily and profitable placered by hand methods with a few years-

Yankee and Ninety-four were almost ghost towns by 1870. The Alice deposit was ground sluiced in 1882 and hydraulicked in 1883. The outcrop material there was three to six feet of rocks and soil above two feet of yellow earth that was in turn above about three feet of white and gray, clayey shattered bedrock containing quartz crystals and pyrite. The yellow earth reportedly "gave a nice string of colors in the pan." The vein outcrops that were placered on Yankee Hill were eluvial placers. The outcrop of the Alice drposit was also an eluvial placer, but the placers that continued for a short distance downslope were colluvial placers.

Although these placers have long since been worked out, careful prospecting might still yeild some "dirt" for panning. ***Be careful around the old mine workings and the edges of the Alice pit and don't enter the workings!***

NORTH EMPIRE DISTRICT

The "Silver Mountain ore zone" lies in the W sec. 21, T. 3 S., R. 74 W., on the west side of North Empire C reek and about one mile north of the U. S. Highway 40. Here there is a large zone of shattered rock, strongly mineralized with pyrite and gold-bearing chalcopyrite. The outcrop material was thoroughly weathered, the pyrite and chalcopyrite leached away and the gold liberated, forming an eluvial placer over the ore zone. There were colluvial placers below it and stream placers on North Empire andEmpire Creek.s These continued to Empire but persisted no further downstream.

North Empire Creek drainage was not glaciated. Empire Creek (below the ore zone) was covered by the Bull Lake glacier that came down Clear Creek valley but not by the pinedale glacier. So these placers developed over long periods, but any placer deposits there may have been at Empire and downstream were swept away.

The North Empire placers were discovered in 1862 and worked for three or four years thereafter.

COSTILLA COUNTY

The placers of the Grayback Mining District, north of Russell in T. 28 S., R. 71 W. (unsurveyed), are probably the first worked in the State. There is a dredge cut 2,200 feet long on Placer Creek in Sec. 36, its downstream end about a mile north of Russell. Placer Creek and its terrace gravels were worked from place to place from Giant Gulch to Strawberry creek in sec. 25 and 26. Officers Bar, the site of the first mining activity, is a large terrace remnant between Grayback and Strawberry Gulches, on the east side of Placer Creek in NE sec. 26. There are a few small placer workings in Grayback Gulch from its mouth on Placer Creek in sec. 26 to its junction with Buckskin Gulch in sec. 13, and in Spanish and Giant Gulches. There is gold-bearing gravel in Buckskin Gulch. Placers have also been found and worked from place to place along Willow Creek for two miles up from its junction with Sangre de Cristo Creek in sec. 18, 19 and 30 T. 28 S., R. 70. W.

In the 1850's soldiers from Fort Massachusetts worked Officers Bar in a small way. The Fort, which was on Ute Creek about six miles north of the present site of Fort Garland, was built in 1852 and abandoned for Fort Garland in 1858. Chinese miners are said to have reworked Officers Bar but there are no records of subsequent placering until 1876 and thereafter not until 1896. Beginning the latter year there was considerable small-scale activity for a few years. A small dredge operated in 1910 and in 1911, until it sank. During the 1930's a few individuals placered here.

In the southwest part of the district there are indurated high gravel deposits, called the Grayback Wash in irregular large remnants from 250 to 300 feet above the level of Placer Creek. The bodies of Grayback Wash on the west side of Placer Creek contain no gold, but one on the east side, which extends from Spanish Gulch to the head of Giant Gulch is gold-bearing, reportedly carrying 3 to 10 coarse colors per pan. The gold ranges from coarse to fine and is commonly scaly. The gravels of Spanish Basin, on Spanish Creek, are derived from Grayback Wash.

Lower gravel terraces are conspicuous in Placer Creek valley and both branches of Middle Creek. From Officers Bar downstream to Mill Creek there are terraces at two levels. 50 and 90 feet above the creek. The terraces upstream from Officers Bar on Placer Creek are not goldbearing. From Middle Creek downstream the terraces on the west side of Placer Creek valley did no pay to work. The terraces on the east side of the valley are goldbearing. Officers Bar was the richest and most extensive of them. These terrace gravels and those of the stream floor were worked from place to place from Officers Bar to the mouth of Giant Gulch. The gold produced ranged from flour gold to round flakes one-eighth-inch across, with occasional nuggets to one-tenth ounce.

Most of the gold was found in gravels within a few feet of bedrock. There is abundant magnetite in the gravels. The gravels of Buckskin Gulch and Willow Creek contain similar gold.

The distribution of the gold-bearing gravels suggests two sources for the gold--low-grade bedrock gold deposits on Grayback Mountain and the Grayback Wash. Of the various remnants of the wash, only the one on the southwest slope of Grayback Mountain contains gold. Consequently, the gold deposits on Grayback Mountain are probably an original source of the gold in the wash as well as the gold in the placers.

Select gravel for panning from any of the places mentioned above. Usually in this district the more sedimentary and volcanic rock fragments a gravel contains, the more likely it is to contain gold. Gravels with predominantly metamorphic fragments are usually barren. Magnetite is not a guide; both barren and gold-bearing gravels contain abundant magnetite. The gravels in many places are covered with a few feet of soil or fine alluvium which is barren; be sure that your sample has some coarse fragments. Along Placer Creek select your samples from the east side. Although the Grayback Wash s indurated, at the foot of its outcrops material weathered from it may accumulate.

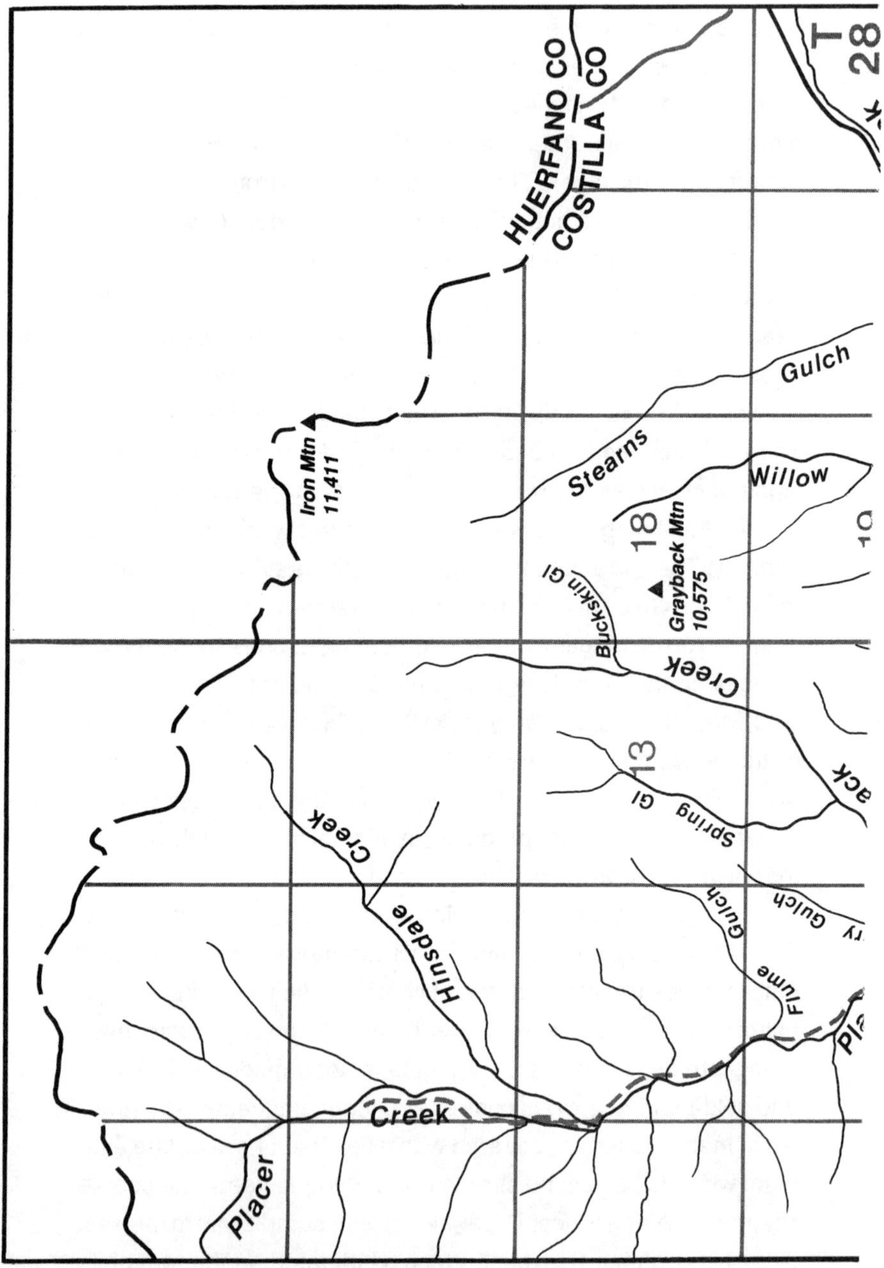
HUERFANO CO
COSTILLA CO
T 28
Stearns Gulch
Willow
Iron Mtn 11,411
Grayback Mtn 10,575
18
13
Buckskin Gl
Creek
Spring Gl
Hinsdale Creek
Gulch
Flume
Placer Creek

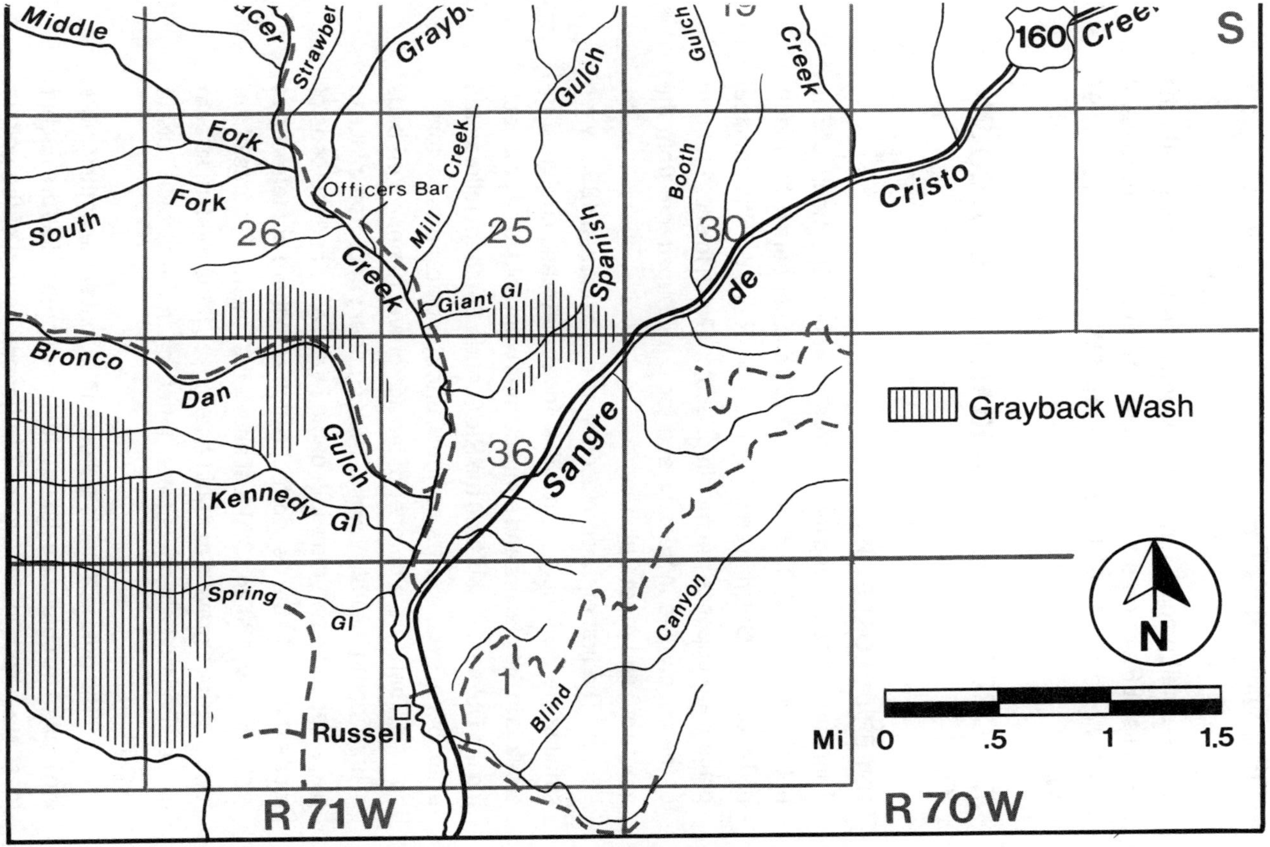

Figure 49. Index of placers in Costilla County. (Modified from U.S.G.S. topographic maps)

DELTA COUNTY

Some small placers have been worked on terraces beside the Gunnison River on the north side of the river near Hotchkiss in southwestern T. 14 S., R 92 W., and in SW sec. 15, T. 15 S, R.97 W.

DOLORES, LA PLATA AND MONTEZUMA COUNTIES

UPPER DOLORES RIVER

There are placers along the Dolores River from the mouth of Burnett Gulch in sec. 2, T. 39 N., R. 11 W., downstream to the mouth of Tenderfoot Gulch in sec. 29, T, 39 N., R. 11 W. The principal placers are opposite the mouth of Scotch Gulch in sec. 11, north of the county line in sec. 15 and for a mile upstream from the mouth of Tenderfoot Gulch.

The first of these placers was discovered in 1882, when small-scale work began. In 1894, for some reason, interest increased in the placers and some 50 men were said to be prospecting the Dolores. At this time the Rio Grande Southern railroad ran a number of excursion trains from Cortez for placer prospectors.

There are a number of terrace gravel remnants on beds of hard rock in the valley walls. Their number varies from place to place depending on the width of the river valley and the beds of rock exposed. In the Tenderfoot Gulch area there are terraces at 40 to 60, 100, 180 and 360 feet above the River, remnants of the lowest terrace gravels are gold-bearing, as it the modern stream gravel. Bedrock channels were drifted on in the terrace cuts in sec. 11.

The gold is a single yellow shade, but some of the colors are rusty. The colors are flakes and wires; some flakes are flat but most are rolled on the edges. The general source of the gold is clearly the Rico mining district. Since very little free gold has been found in the Rico district, perhaps the specific sources are yet to be found.

Select samples for panning from the three principal terrace placer areas already mentioned or from point bars in the river just downstream from them. Take the terrace samples as close to bedrock as possible.

LA PLATA MOUNTAINS

Gold-bearing gravels occur in many places here: 1) along the West Mancos River from the east side of T. 37 N., R. 12 W. to sec. 36, T. 37 N., R. 13 W. 2) along the East Mancos River from its head in Rush Basin to its junction with the Middle River in sec. 24, T. 36 N., R. 13 W. (Gold is also said to have been found from place to place along the Mancos River to its junction with the San Juan River.) 3) along the La Plata River from Bedrock and Boren Creeks in secs. 4, 9, and 10, T. 36 N., R. 11 W. to sec. 10 T. 35 N., R. 11 W. 4) along Junction Creek in southwestern T. 36 N., R 10 W 5. 5) terrace gravels southeast of the junction of the East and Middle Mancos Rivers in sec. 19, 29, and 30, T. 36 N., R 12 W. 6) and northwest of Hesperus in secs. 3, 4 and 10 T. 35 N., R 11 W.

Gold was known on the La Plata River in 1874 and was found on Junction Creek in 1885. In 1887 there was placering on the East Mancos River. This work was not very productive. There were a number of small-scale operations in the 1930's. In 1932 or 1933 placering on the East Mancos River led to the discovery of the rich Red Arrow bedrock gold deposit.

The gold is said to be coarse on Junction Creek but extremely fine elsewhere, so on the various rivers select panning samples from point bars.

BAKER'S BRIDGE

There is a placer deposit upstream from Baker's Bridge in parts of sec. 19 and 30, T. 37 N., R. 8 W. and secs. 24 and 25, T. 37 N., R 9 W. Baker's Bridge is near the west quarter corner of sec. 19, at head of the broad Animas River valley. Near the north quarter corner of sec. 19, at the head of the broad Animas River valley. Near the north quarter corner of sec. 19 is the mouth of the deep gorge of the Animas River. In the gorge and below the bridge the course of the river is southerly. Between the gorge and the bridge the river's course is southwesterly for nearly half a mile through a 1,000 foot wide, park-like valley to the west side of sec. 19. Here it enters a short 50 foot deep canyon just above the bridge.

Although these placers have been known for many years, they have not been much worked. The Baker party was camped near here in the summer of 1961; Baker's Bridge is name for him. It appears that the party did not find this place.

Between Baker's Bridge and the mouth of the gorge there are at least three abandoned channels cut in the granite bedrock. Two of these are beneath terrace gravels; they were found by drilling. The highest, terrace 180 feet above river lever is marked by a few gravel remnants on the bare granite bedrock. In the park there are three lower terraces. The highest of these is 70 feet above the river.

The high terrace remnants, the lower terrace gravels and the gravels on the present bars and banks of the river are all more or less gold-bearing. Most of the gold here is fine, but some coarse gold and a few small nuggets have been recovered. The gold is of several colors--yellow, red-orange and distinctly greenish--reflecting many different bedrock sources upstream.

Most of the gold is bright, but some has rusty surfaces. It averages 817 fine.

There was a Pinedale recessional moraine at Baker's Bridge, damming the valley there. It forced the meltwater stream of that time to flow high on the east side of the valley, where it cut the channel of the high terrace. After the moraine dam was considerably eroded, the meltwater stream moved westerly and downslope, successively cutting the abandoned channels now buried beneath the lower terraces and finally the present 50 foot canyon above Baker's Bridge. The lower terraces have been formed as the river has repeatedly filled and then cut again in its present valley floor. The gold came from the various gold camps up the Animas River, both from the erosion of the morainal dam, formed from glacially transported material, and from gold brought from upstream by alluvial transportation which has continued wince the glacier retreated.

Select gold for panning first from the skim bars or the banks. Their gold is fine but the skim bar values here are unusually high. On the terraces even their soil may have a few colors.

EAGLE COUNTY

Terrace gravels of the Colorado River contain gold and have been worked in Eagle and Grand Counties. The principal one in Eagle County is the Gately placer in sec. 3, T. 2S., R. 84 W. These placers were discovered before 1894 and worked sporadically until 1942.

The river's canyon is cut largely in alternating sandstones and shales--hard and soft rocks. In many places the river has stripped away soft rocks exposing resistant beds, which form benches along the river. The terrace gravels lie on these benches. The terrace gravels range up to 50 feet thick; gold is concentrated in the gravels from bedrock to seven feet above. The gold is fine-grained and difficult to recover. There is much black sand.

LAKE CREEK

Placers have been worked on Lake Creek near its mouth in sec. 6 and upstream in sec. 7 and 18 and in Creamery and South Creamery Gulches near their junction in sec. 8, T. 5 S., R 82 W., The placers were discovered in the 1870's. They are in outwash of the Lake Creek glaciers below an end moraine in sec. 18, T. 5S., R 82 W. The grade of the gravel is reportedly good. A little sporadic placering from 1932 to 1941 is reported in the Mt. Egley district, perhaps in sec. 21 and 22, T. 6 S., R 82 W. In this high,

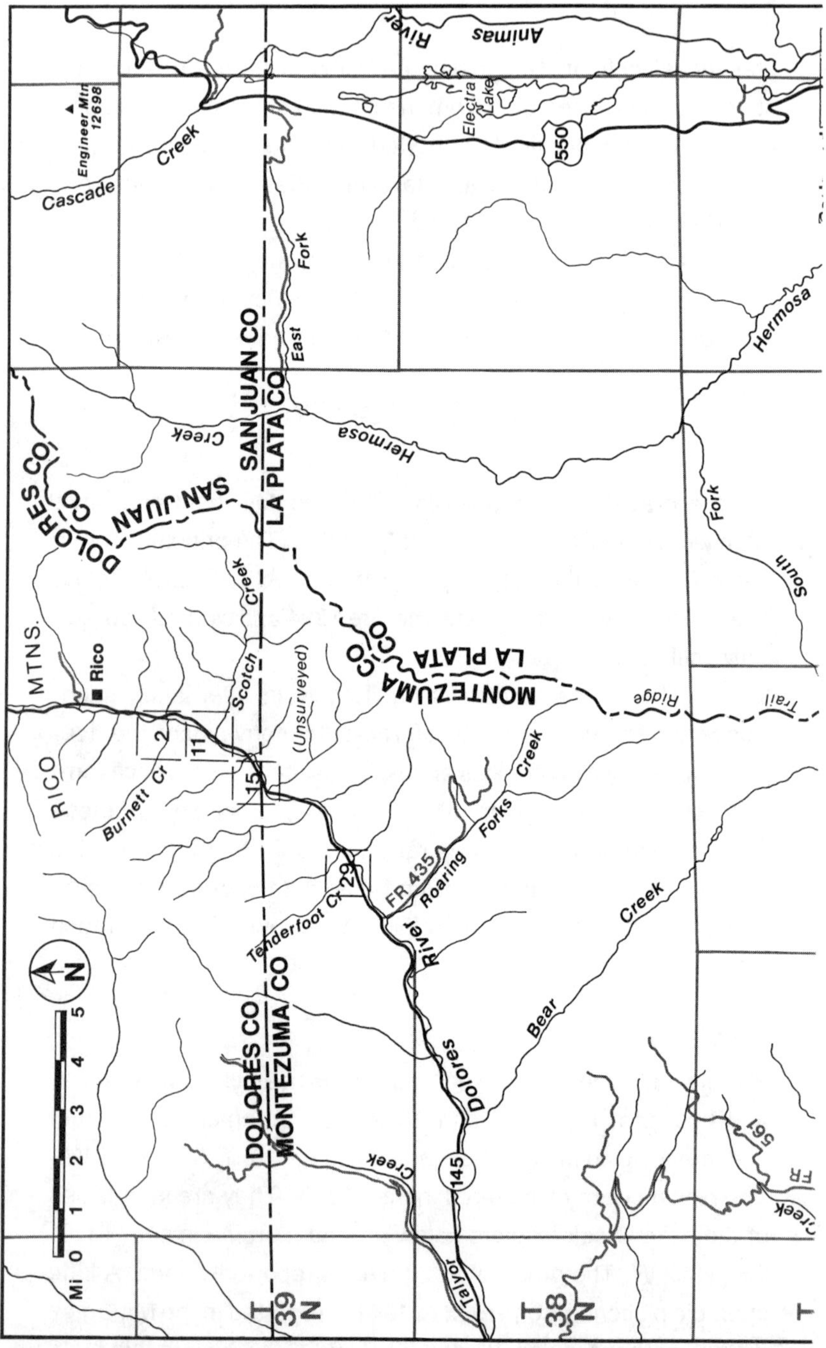

Engineer Mtn
12698
Cascade
Creek
Animas
River
Electra
Lake
550
East
Fork
Hermosa
Creek
Hermosa
South
Fork
DOLORES CO
SAN JUAN CO
SAN JUAN
SAN JUAN CO
LA PLATA CO
MONTEZUMA CO
LA PLATA CO
Ridge
Trail
RICO
MTNS.
Rico
Scotch
Creek
(Unsurveyed)
Burnett Cr
2
11
15
29
Tenderfoot Cr
FR 435
Roaring
Forks
Creek
River
Dolores
Bear
Creek
DOLORES CO
MONTEZUMA CO
Creek
145
Talyor
561
FR
Creek
Mi 0 1 2 3 4 5
N
T 39 N
T 38 N
T

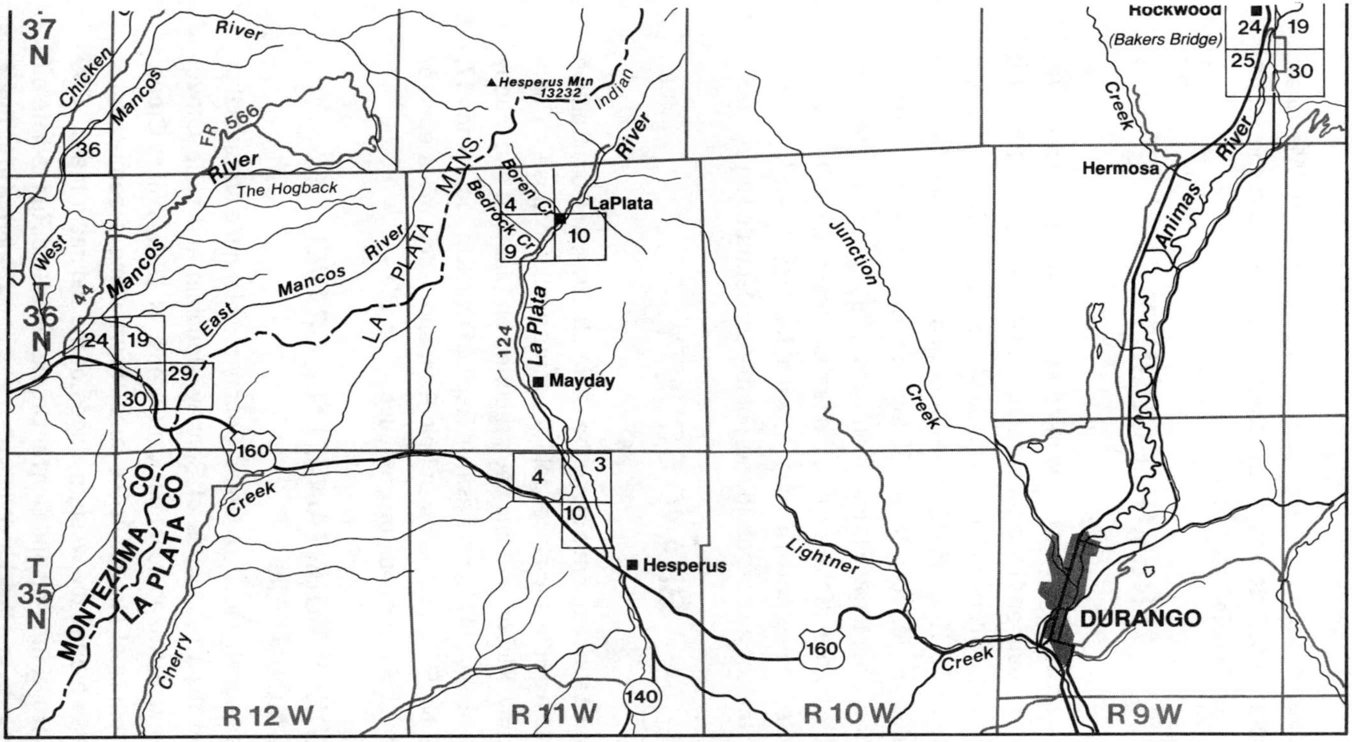

Figure 50. Index map of Upper Dolores River, La Plata Mountains, and Baker's Bridge placers in Dolores, La Plata, and Montezuma Counties. (Modified from U.S.G.S. topographic maps)

upstream area any placer deposit is probably post-Wisconsin, similar to the placers on the South Platte River at the foot of Hoosier Pass in Park County.

FREMONT COUNTY

Small placers have been worked at several places on the Arkansas River and its lower terraces in Fremont County. Typical of them are placers on the north side of the River at Texas Creek in sec. 13, T. 19 S., R. 12 E. Here for about half a mile there are working on terraces 16 and 30 feet above the river level. Gravel on the lower terrace is less than 10 feet thick. The cuts there are within 20 feet of the edge of the terrace; few of them have been worked to bedrock. This suggests that the gravels are low-grade and that the gold is fine and not notably concentrated on bedrock.

GRAND COUNTY

Some terrace gravels along the Colorado River in Gore Canyon contain a little fine placer gold. They are similar to those in Eagle County. Placer gold has been reported on Diamond and Muddy Creeks east of Muddy Pass, on Red Dirt Creek in sec. 16 and 17, T. 3 N., R 82 W., below Buffalo Park and "a few miles east" of Grand Lake, probably on East Inlet.

BENNAY MOUNTAIN-LITTLE GRAVEL MOUNTAIN

The largest placers of Grand County are along the streams that drain the west slopes of Bennay Mountain and Little Gravel Mountain, two adjacent peaks in T. 4 N., R 77 W. Willow Creek, the principal stream in the area, flows along a semi-circular course around the western slopes of the mountain mass. Elk, Bronco, Kauffman and Denver Creeks and Gold Run radiate off the peaks and flow into Willow Creek. Gold is reported in all these creeks and some placering has been done. No gold has been found in Willow Creek's tributaries from the north and west. The largest placers were those on Gold Run from its mouth for about four miles upstream. Gold has been reported also on Stillwater Creek, which drains the east slopes of the mountain mass.

These placers were known by 1871, and they have been worked occasionally since. Their production has been small.

On the summits of the two mountains there are deposits of coarse, bouldery gravel over 200 feet thick. The stream placers were probably formed by reworking these high gravels, since no bedrock gold deposits are known in the Will Creek or Stillwater Creek watersheds. The gold in the stream placers is fine; their gravels are sandy and easily washed and contain much black sand. At Gold Run there are two terraces, 40 and 75 feet above the stream. Only the stream and lower terrace gravels have been worked.

In this area to begin select samples from Gold Run, Kauffman Creek and along Willow Creek, below the mouths of these two creeks.

GUNNISON COUNTY

GOLD BRICK DISTRICT

Placers were worked on Dutch Flats Gulch, Spring Gulch and Jones Gulch in sec. 12, T. 50 N., R. 3 E., secs. 1 and 12, T. 50 N., R. 3 E., and secs 6 and 7, T. 50 N., R. 4 E. Dutch Flats Gulch is four miles and Jones Gulch is three miles north of Ohio City; Jones Gulch lies between them.

The story of these placers' discovery in 1861 is told in the Tincup Gulch section below. Apparently they were worked out in a few years.

The placers on Dutch Flats Gulch were in Dutch Flats, a more gently sloping park in the upper part of the Gulch. The placers were shallow, with two to three feet of gravel beneath up to five feet of overburden. There are working for 700 feet across the Flats, widening from 15 to 50 feet wide at their eastern end. At their eastern end they appear to be near their source.

TINCUP GULCH

Tincup Gulch is in sec. 19, T. 15 S., R 81 W., (Unsurveyed). It is a hanging valley tributary of Middle Willow Creek. (The Middle Willow creek glacier, with a large, high ice-gathering area, scoured the Willow Creek valley below the level of the floor of its smaller tributary, Tincup Gulch.) The upper section of Tincup Gulch, eroded by Tincup Creek, is separated from the Middle Willow Creek valley floor by a steep section where Tincup Creek plunges down the side of Middle Willow Creek valley.

The first discovery of placer gold in Gunnison County was in Tincup Gulch. James Taylor, for whom the river was named, set out from Granite in 1860 with three others to prospect west of the Divide. Tincup Gulch and the town get their name from the tin cup Taylor is supposed to have used to pan when the discovery was made that summer.

Taylor and his party returned to Tincup in 1861. Three men followed them, hoping to learn the where-abouts of the Tincup discovery. These three lost the Taylor Party's trail, wandered in the mountains and found placer gold in Dutch Flats, in the Gold Brick District, and in Union Park, both mentioned elsewhere. (The three would-be trackers were Karl and Fred Seigul and Fred Lottis. The Sieguls were remembered in "Dutch" Flats and Lottis Creek in Union Park is named for Fred Lottis.)

Tincup Gulch was worked by hand until 1885, at least, when one of the original Taylor party was reported working there.

The placers of Tincup Gulch are all in the upper section of the Gulch, above Middle Willow Creek valley. They are found from place to place in the lower half mile of this upper section; in this distance the workings were almost continuous the last 1,000 feet. The gravels were shallow and all the placers were worked to bedrock. The gold was derived from the important Gold Cup ore deposit, which crops out on the east slope of the Gulch, and some small veins on the southwest side of the ridge.

WASHINGTON GULCH

Washington Gulch is a left-hand tributary of the Slate River in the Western half of T. 13 S., R 86 W.

Washington Gulch was discovered in 1860, worked for a few years and then abandoned. The machine placering described below was done after 1947 and the drift mines described showed signs of recent work in 1958. In the early days Washington Gulch was well known for its high grade gravels. One pioneer said that 5 to 10 ounces was the average day's production, but he once had a pan that yielded $78 (probably, allowing for fineness, more than four ounces),

Washington Gulch has three sections; the upper cirque in which the abandoned town of Elkton was located, the lower section with a wide valley floor and between the two a deep, steep walled gorge about a quarter of a mile long. There are machine placers below the gorge for a quarter of a mile and then old small cuts from place to place through the SE sec. 8. The gorge section has been machine placered. At the head of the gorge there is a hydraulic cut in glacial moraine there and small, creek-bed placers for a quarter of a mile upstream. There are no creek-bed placers above this, but a slope from the west here has been ground-sluiced for 360 feet from the stream. At the top of the sluice cut there are a number of shallow drift workings The gold was coarse. Its source has not been recognized.

Search for gravel to pan throughout the old placered sections. Some tailings from past operations should still contain a few colors. Unworked, lower-grade gravels can probably be found in the moraine at the head of the gorge and along the stream in sec. 9. The moraine has rocks of all sizes, mostly angular, in a very clayey matrix. In SE sec. 8, the old miners apparently worked only near-surface stream bed gravels; except for an occasional check pan, you can limit your panning to the same.

WILLOW CREEK

The surface gravels of Willow Creek were worked by hand in the early days from the mouth of Bertha Gulch (sec. 6, T. 15 S., R 81 W., Unsurveyed) for about 1.1 miles downstream and dredged in 1913 from 0.6 to 1.1 miles downstream from the mouth of the Gulch. Reportedly some work was done upstream at the mouth of Tincup Gulch but I have not been able to find the workings. Probably some colors can be found along Willow Creek from the mouth of Tincup Gulch (sec. 20, T. 15 S., R 81 W., Unsurveyed) to the Taylor River Reservoir.

HUERFANO AND LAS ANIMAS COUNTIES

A Little place gold has been produced on Wayatoya Creek on the north slope of West Spanish Peak in Sec. 26, T. 30 S., R 68 W., a short distance upstream from the end of the road. Reportedly, gold-bearing gravels occur near timberline on several gulches on the east slope of East Spanish Peak and on the saddle between the two peaks.

JACKSON COUNTY

In Jackson County the principal placer workings are at Independence Mountain. Some very limited placer activity reportedly took place from time to time in the southwestern part of the county, apparently on Grizzly Creek in the northwestern part of T. 5 N., R. 81 W., and the northeastern part of T. 5 N., R. 82 W. on the northwest slopes of the Rabbit Ears Range. Placer gold is also reported in northwestern Grand County, across the range on Diamond and Muddy Creeks. The gold on Grizzly Creek is said to be finer than in the creeks across the Range. It is difficult to find these occurrences; careful prospecting and digging to bedrock are necessary. The location of these showings suggests possible sources in the Rabbit Ears Range east of Rabbit Ears and Muddy

Passes.

INDEPENDENCE MOUNTAIN

The Independence Mountain placers lie on the southwest slopes of Independence Mountain in secs. 23 (along Threemile Draw), in secs. 26, 35 (along Placer Draw) and along unnamed gulches in sec. 36 T. 11 N., R 81 W., and in sec. 31, T. 11 N., R. 80 W. Placers were continuously worked here after the completion in 1897 of a 19 miles-long ditch and pipeline from Big Creek, which lies to the northeast of the placers.

The mountain is 9,726 feet high. Its southern side, below about 8,800 feet elevation, slopes rather gently toward North Park. Placer Draw flows in a 100 foot deep valley cut in this surface. Above about 9,000 feet elecation Independence Mountain is covered with coarse, bouldery gravel. The gravels of the streams, as well as the gravel and boulders lying on the lower slopes,are derived from this gravel cap. No bedrock gold deposits are known on Independence Mountain. The mountain-top gravels have been proposed as the immediate source of the gold in the placers, but this explanation does not account for the absence of placers in other gulches which head in the Independence Mountain gravels.

Bedrock is a soft sandstone in which some stream channels have been incised. The two placer cuts extending from the east side of the main cut in placer Draw followed channels different from the modern stream courses. The gravels in Placer Draw pinch out quickly to either side of the draw, except at the side channels just mentioned. They are nearly mined out, although a little gravel remains at the head of the cut. The gravel continues down-stream from the cut, but could not be worked because the gradient was too gentle to wash the gravel from the cut with the water available. The gravels on Placer Draw are sandy and easily washed, but those on Threemile Draw are clayey.

LAKE COUNTY

LEADVILLE DISTRICT

In a few places Terrace No. 1 gravels, glacial outwash and the gravels of the Arkansas River valley floor have been tested or worked. Placered gravels in Terrace No. 1 can be found on the south side of the draw near the center of sec. 21, T. 9 S., R. 80 W., and south of Yak Tailings pile in a large hydraulic cut in N. sec. 25, T. 9 S., R 80 W. Meltwater channels of the East Fork moraine have been worked in cuts on the south side of the river in NW sec. 14, T. 9 S., R 80 W. Outwash of the Evans Gulch glacier was worked at the Pearley placer in the gulch in S sec. 21, t. 9 S., R. 80 W. Outwash deposits have been worked in two small cuts along Iowa Creek in sec. 32 and 33, T. 9 S., R 79 W. The gravels of the Arkansas River valley floor have been extensively tested and are gold-bearing; the gold is near bedrock and inaccessible. There may be a little fine gold in the river bar gravels.

BUCKEYE GULCH

Buckeye Gulch has formed an alluvial fan where it flows from the high mountains onto the valley of the East Fork in sec. 32, T. 8 S., R. 79 W. At the top of the fan, just below the mouth of the hanging, mountain valley, a placer cut exposes porphyry bedrock beneath reworked till. Select panning gravel from the head of the cut near bedrock or from the face of the lower end of the cut. Lower on the fan the surface might be prospected for other channels. This is an alluvial fan deposit of gold from small veins on the south face of Buckeye Peak.

EAST TENNESSEE CREEK

East Tennessee Creek and Thayer Gulch were not glaciated. Alluvial placers were worked on the former in secs. 23-27, T. 8 S., R. 80 W., and on the later in secs. 14 and 15. The source of their gold is the veins of the Buckeye Peak area. There is some unworked gravel remaining at the Thayer Gulch cut and in isolated patches in the East Tennessee placered area. The gravel downstream is too deep to reach.

COLORADO AND LITTLE FRYINGPAN GULCHES

There are extensive placers in lower Colorado Gulch from its junction with the Lake Fork to Little Fryingpan Creek, a tributary of Colorado Gulch from the south slope of Sugarloaf Mountain. These are in sec. 30 and 31, T. 9 S., R. 80 W., and sec 25, T. 9 S., R. 81 W. Reportedly, these placers were rich. They are stream alluvial placers with gold derived from the low-grade veins of Sugarloaf Mountain. These valleys were not glaciated; the formation of the placers began in pre-glacial times and has continued since. The valleys have been extensively placered and only a little gravel remains in isolated patches. If you find an unworked patch, clean the weathered bedrock as deeply as you can; it is an excellent trap.

LAKE CREEK AND FLUME GULCH

The only placers in Lake Creek valley itself are those near the mouth of Lake Creek in sec. 24, T. 11 S., R. 80 W., and the small placer in Flume Gulch to the south of Twin Lakes in Sec. 29 (unsurveyed). Along the Arkansas there are placers on outwash terraces overlooking the river in secs. 11, 13, 24, 25, and 36, T. 11 S., R. 80 W. The Lake Creek placers were discovered in 1860 and worked on a small scale for a few years after that. From 1898 until 1908 the placers were tested and some hydraulicking and drifting done.

The Twin Lakes are glacial lakes: the west lake dammed by the upper (younger) Pinedale terminal moraine of the Lake Creek glacier and the east lake dammed by the lower (older) Pinedale and Bull Lake moraines. Lake Creek has cut its valley through the eastern moraines; it is floored with lower Pinedale outwash. Two gulches to the north of the valley mouth that have been hydraulicked-- New York and Young American Gulches--

are floored with upper Bull Lake outwash gravels. These are the largest of the placer cuts in the outwash terraces mentioned above.

The Lake Creek gravels here have been tested and have good values. Probably they were never worked partly because of the large boulders they contain. The boulders increase in size upstream as the moraine is approached; near the moraine 15 foot boulders are not uncommon. Test shafts have shown that the boulders tend to decrease in size and in number with depth. This suggests that glacial till and outwash may have been deposited upon older alluvial gravels. Although most of the gold occurs near bedrock, it is found in ":streaks" --high grade layers separated by leaner or barren layers. The gravels are clayey and the gold is coarse. The higher, older gravels exposed in the New York, Young America and other terrace placers are similar. Meltwater streams reworking the moraine on the south side of the valley formed the Flume Gulch placer.

The immediate source of the gold is the moraines below the Twin Lakes. The original source is the many gold veins in the Lake Creek watershed, only a few of which have been rich or large enough to work individually. Expect no important placer deposits in the valleys between the Twin Lakes and these source veins; they were all destroyed during the glaciation.

Select samples wherever gravels are exposed. Coarser gravel sections, which might represent channels on false bedrock, are the most favorable. Since the gold occurs in streaks, you may have to wash gravels from several levels and places before getting colors. At some places in the old hydraulic cuts, gold can be found in varying amounts from the cut floors to near surface.

TWOBIT AND SAWMILL GULCHES

There are placers in Twobit Gulch for about half a mile upstream from the south line of sec. 1, T. 11 S., R. 80 W. and in Sawmill Gulch for about a quarter of a mile upstream from the center of sec. 2. Some placering has also been done on the Arkansas River at the mouth of Twobit Gulch in sec. 11. Here gray sand "follows the gold" (occurs in the heavy minerals in the pan concentrate). This gray sand is probably lead minerals from the Twobit Gulch drainage.

LAKE AND CHAFFEE COUNTIES

ARKANSAS RIVER AND TERRACES ALONG IT

Placers have been worked in the bed of the Arkansas River and in elevated terraces along it from just above the mouth of Lake Creek downstream to the mouth of Railroad Gulch in Browns Canyon, 34 miles to the south. The stream bed placers were discovered in 1859 and 1860. Within a few years they became "winter diggings" where miners could at least pay for their grub when higher altitude diggings froze. There have been small-scale operations intermittently since, especially during the 1930's.

Bedrock beneath the valley floor has never been worked, although it has been reached in a few sampling shafts. In 1861 and again in 1877 attempts were made to flume the Arkansas to permit working the deep gravels and in 1927 a suction dredge was built to work them. These attempts failed. Some workings on the terraces were probably hydraulicked, but most were worked by hand. The gravels on the Arkansas River have been worked: near the moth of Twobit Gulch (just described), from the county line downstream to about half a mile below Granite, from the mouth of Clear Creek downstream for about a mile, at the mouth of Cottonwood Creek at Buena Vista, southeast of Nathrop and at the mouth of Railroad Gulch near Brown's Canyon. One place early miners considered particularly good was Georgia Bar, on the west side of the Arkansas at the mouth of Clear Creek, in SENE sec. 8, T. 12 S., R. 79 W.

I have seen gold washed in 1958 just below Georgia Bar. It was very fine and flaky. The miners said that this fine gold made a fine tail in a pan but was very difficult to recover. It was probably skim bar gold It was of several colors, which might be expected considering that the gold probably came from several of the various gold-bearing districts upstream.

Terrace gravels have been worked: from the mouth of Lake Creek southward for about one mile on the west side and much less on the east side of the river; near the center of sec. 5, T. 12 S., R. 79 W., southwest of Granite; on the east side of the River at Georgia Bar and south of Princeton in sec. 22, T. 12 S., R. 79 W. The terraces are approximately 20 and 30 to 50 feet above the valley floor.

MESA COUNTY

Some placer gold has been produced in Mesa County from the Colorado and Gunnison Rivers and in Unaweep Canyon. The production probably was from skim bars in the modern streams and from elevated terraces along the rivers. The location of these placers is unknown.

MOFFAT COUNTY

The principal placers in Moffat County are those of the Iron Springs Divide area, described below. In addition, there are placers gravels on the Yampa, Little Snake and Green Rivers. They are low-grade and little work has been done on them. On the Yampa they are at Round Bottom, on the north side of the River in sec. 34, T. 6 N., R. 92 W., and also in terraces 20 to 50 feet above the River, covered with up to 30 feet of silt, on the south side of the River on the south side of Big Bottom, in T. 6 N., R. 91 W.

On the Yampa and Little Snake Rivers some very fine gold has been produced from bars downstream from the junctions of Lay and Fortification Creeks with the Yampa and from the junctions of Fourmile Creek an Dry, Scandinavian and Bighole Gulches with the Little Snake. (All these streams head in the Iron Springs Divide.) On the Green River, placer gravels are reported near Brown's Park in the northwestern part of the county. (Brown's Park is in T. 10 N., R. 103 and 103 W.)

IRON SPRINGS DIVIDE

Iron Springs Divide is a broad, sinuous ridge which extends from Black Mountain in T. 10 N., R. 90 W., in a westerly direction to the northern park of T. 9 N., R. 93 W., between Great Divide and Iron Springs. To the north of the Divide, the gulches drain into the Little Snake River; to the South, the gulches drain ultimately into the Yampa River.

The placers lie from place to place within a large area in T. 8 to 12 N., R 91 to 94 W. Gulches in which placer gold has been found on the north slope of the divide are lower Fourmile, Timberlake and Housel Creeks and Pole, Dry, Scandinavian and Bighole Gulches. One the south slope of the Divide gold has been found in Lay Creek and Dressler Gulch and in Blue Gravel Creek. Over much of the divide area, most pansful of soil taken will show at least one color.

The first reported placer gold discovery in this area was on Fortification Creek in 1882. Apparently the find was forgotten, and there was no activity for nine years. In 1891 Hugh Morrison, a prospector from Aspen on his way to the Uintah Mountains, found gold on Fourmile Creek. He prospected and found that this gold did not come from the Elkhead Mountains to the east. He then prospected along the Little Snake Ricer and apparently found gold on it and at the mouths of Dry, Scandinavian and Bighole Gulches. He went to Aspen in December and returned to the Iron Springs Divide country in February the following year. In 1892 he located claims at his four discoveries and built a cabin and began placering on Fourmile Creek. Operations by individuals continued each year for twenty years or more, and revived during the

Depression. There had been several pulses of activity since World War II, but none of them has proved profitable

In 1893 hydraulicking began on the Fourmile. Between 1895 and 1903 several ditches were completed to deliver water to the northern gulches. In 1900 B. L. Law built a small dredge on Timberlake Creek in sec. 8, T. 11 N., R. 91 W. It was operated successfully until 1904, when it burned. In 1905 another small dredge, the "Blevins dredge," was built on Lay Creek and worked for several seasons in secs. 15 and 16, T. 8 N., R. 93 W. In 1937 to 1939 a dry-land washing plant was operated on bars above Timberlake Creek in secs. 2 and 11, T. 10 N., R. 92 W. These last three operations recovered 26 to 50 cents worth of gold per cubic yard, at the prices prevailing at the times of production. Reportedly, only the top three to six feet of the gravel was mined each time. Apparently at most placed they did not mine to bedrock.

Tertiary sedimentary rocks form the bedrock throughout the placer area. They are composed of shales, sandstones and several pebble conglomerates; most of these are soft and poorly consolidated. The placer "gravels: are sands, predominantly of quartz, with a few pebbles, usually less than one-inch diameter. Some larger pebbles and a few cobbles occur, but they are very rare. At many places throughout the district the placer "gravels" are very much like the weathered bedrock and it is difficult to distinguish them.

Although a trace of gold can be found in the gravel at most places over the divide, gold content abruptly diminishes above about 7,100 feet elevation. Near this elevation drift mines were worked near Iron Springs and at the head of Housel Gulch. They followed channels on a clay false bedrock. This blue Claystone is exposed in Lay Gulch east of Iron Springs. Below the elevation of these drift mines there are many stream and terrace placers as well as slopes with traces of placer gold in the soil. The immediate source of gold of these deposits is fossil placer channels in the bedrock. A number of such channels probably are exposed beneath the mantle near the 7,100 foot level on the divide, and because of the wide distribution of the placers below and the presence of a little gold above that level, there probably are buried channels at other levels.

Along upper Timberlake and Lay Creeks and many of the gulches there are terraces at two levels. Both the stream beds and the terraces have been worked at many places. The most extensive workings are on Timberlake and Lay Creeks. The Law dredge workings with ground sluiced areas on either side are in sec. 5 and 8, T. 11 N., R. 91 W. The dry land plant worked terraces at the junction of Housel and Timberlake Creeks in sec. 2, T. 10 N., R 92 W. The Blevins dredge worked in a terrace in sec. 22, T. 8 N., R 93 W.

The gold from the Divide is very fine-grained--some 67 colors per milligram (2,067,000 per ounce). It is also very pure, ranging from 885 to 935 fine. The colors are rounded rather than flaky.

Although there is said to be "gold in every shovelful" on the Divide, you'll do best beginning prospecting at the edges of the old machine workings on Timberlake and Lay Creeks mentioned above. Pan around one of these workings to find its best side; then select a sample from two to three feet down in unworked ground near that best side.

OURAY COUNTY

Ouray County, in common with the other counties in the San Juan Mountains, has no placer deposits corresponding in size to the important gold mineralization in its watersheds. A few fine colors, post-glacial and stamp mill gold, can be found along Red Mountain Creek and, especially, Canyon Creek above Ouray. As is true of valley floors in the Silverton and Telluride areas, no economic placer mining was ever done on them.

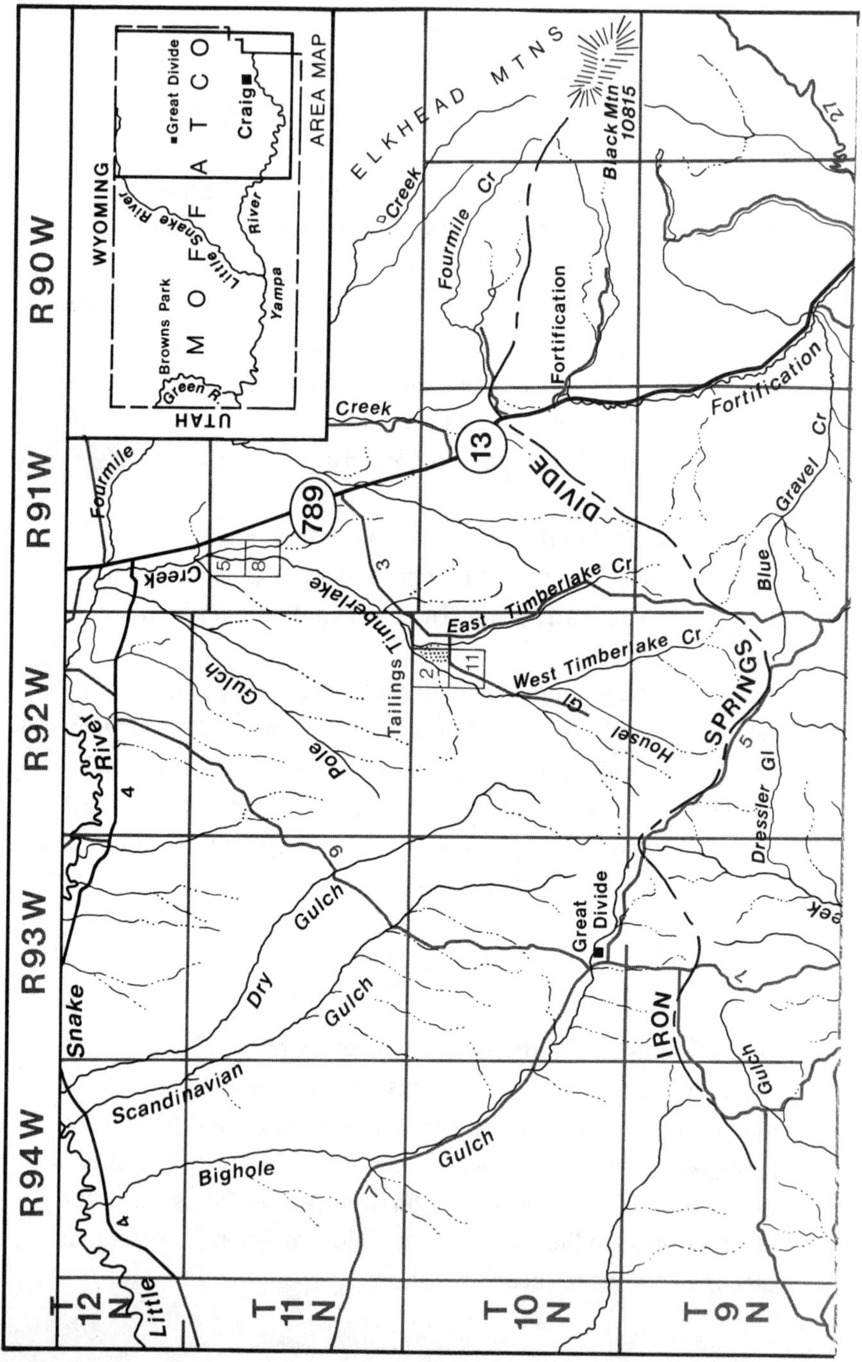
R90W
R91W
R92W
R93W
R94W
T 12 N
T 11 N
T 10 N
T 9 N
WYOMING
UTAH
Green R.
Browns Park
Little Snake River
MOFFAT CO
Great Divide
Craig
Yampa
River
AREA MAP
ELKHEAD MTNS
Black Mtn 10815
Creek
Fourmile Cr
Fortification
Fortification
Creek
13
789
DIVIDE
Fourmile
Creek
Timberlake
East Timberlake Cr
Gravel Cr
Blue
Tailings
West Timberlake Cr
Housel Gl
SPRINGS
Gulch
Pole
Snake River
Dry Gulch
Great Divide
Dressler Gl
IRON
Gulch
Gulch
Scandinavian
Bighole
Gulch
Little

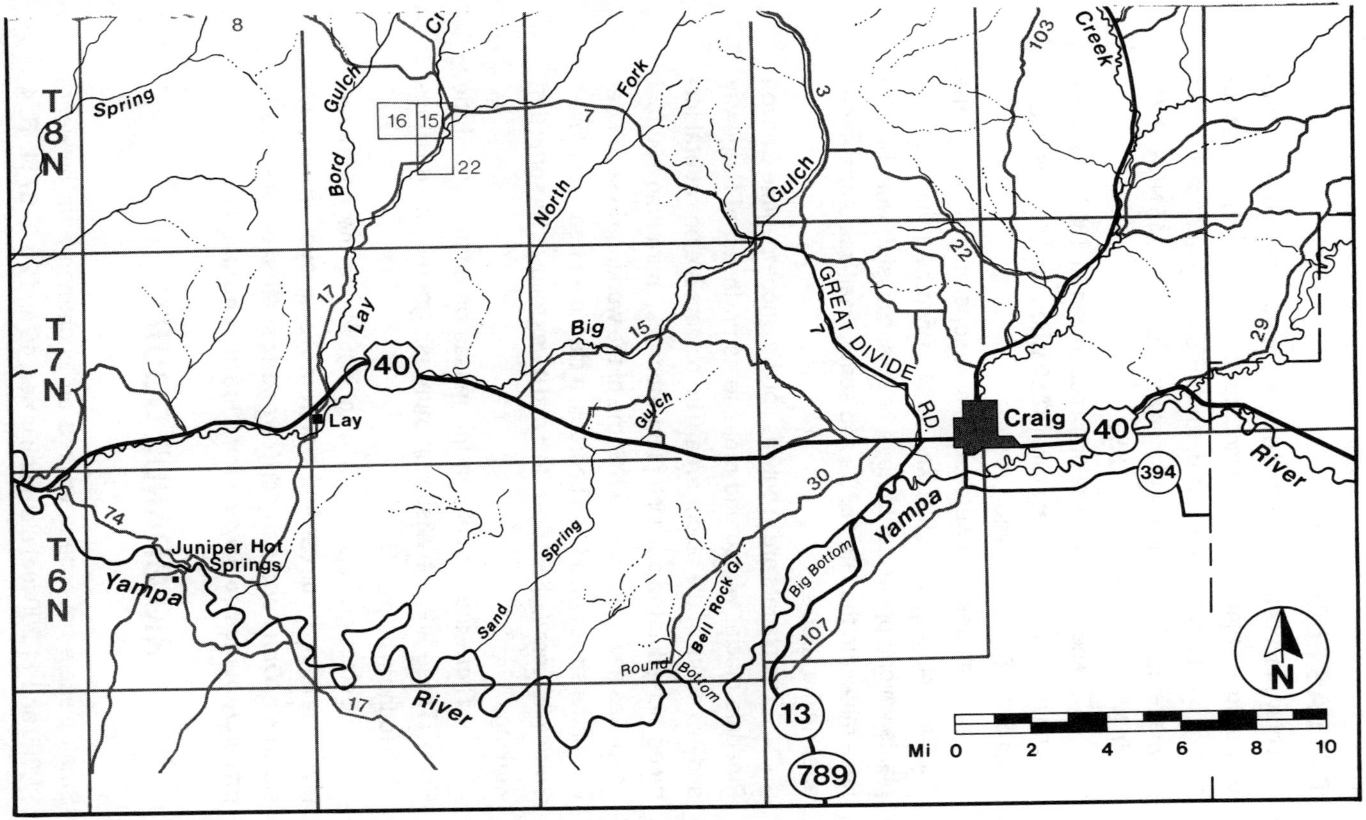

Figure 51. Index map of Iron Springs Divide placers, Moffat County. (Modified from U.S.G.S. topographic maps)

DALLAS PLACERS

The only placers that have had development of any consequence at the Dallas placers on the Uncompaghre River, which extends from a point about two miles north of Ridgeway almost six miles downstream. The largest workings are in sec. 4, T. 45 N., R. 8 W., and in NE sec. 17, T. 46 N., R. 8 W.

These placers were first recognized in 1860, but there was no activity for 20 years, until 1880 when the Ute tribe ceded most of the San Juan region.

Activities here were; small-scale operations Through the 1880's, installation in 1883 and 1884 of two hydraulic elevator plants which appear not to have been successful, work by Chinese miners in the late 1890's and some small-scale operations again in the 1930's.

The placers are in outwash of the Uncompaghne Canyon Pinedale glacier whose end moraine is northeast of Ridgeway and just south of the placer area. The placers are on several terraces incised in the outwash plain. Bedrock, more than 100 feet deep, has never been tested. Some of the workings in sec. 4 developed a streak on a false bedrock, a clay layer near river level. Some pits were worked down to it and it was followed in some drift workings.

There are gravels on the mesas on either side of the River. These are pre-Wisconsin outwash gravels; no gold has been found in them.

Select gravels for panning just above the near-river-level clay layer, if you can find it, or from coarse layers in the gravels above. ***Don't enter any drift workings! Most of this area is privately owned so permission should be obtained.***

RIO GRANDE COUNTY

Small placers have been worked on Wightman Fork between 11,000 and 11,200 feet elevation in sec. 19 and 20, T. 37 N., R. 4

E., near Summitville.

These placers have been worked on a small scale from time to time from their discovery through the Depression times. Although their output has been small, the placers have yielded many nuggets. The largest, found in 1880, weighed more than 17 ounces and one to four-ounce nuggets reportedly were not uncommon.

The ground placered in the past is on the floor of a broad glacial ampitheater above the head of the gorge section on Wightman Creek in sec. 20. The placering was in the stream bed and on the slopes below the veins.

Most of the ground near Summitville is private property. Ger permission before you start panning!

ROUTT COUNTY

HAHNS PEAK

In the Hahns Peak District there are two principal placer areas: Poverty Bar on deep Creek northwest of Hahns Peak village insec. 20 T. 10 N., R. 85 W., and Ways Gulch to the east of the village. There are smalled placers in Little Red Park on a branch of King Solomon Creek north of Twin Mountain and in the vicinity of Columbine. A little mining and some testing has been done along Willow Creek.

The Hahns Peak placer area was discovered by a Captain Way in 1864. In Empire he told John Hahn of his discovery. In 1865 Hahn and W. A. Doyle went to the area and prospected it. They returned with 40 men in 1866, built cabins at the site of the present village and placered through the summer. They established the mining district and named it and the peak for Hahn. Facing an early winter, the 40 men returned to Empire, leaving Hahn and Doyle at the camp. In the spring oof 1867 these two needed supplies. They headed for Empire, and in a blizzard during their journey Hahn died.

Although the district has been only a modest producer, it supported some activities each year from 1866 through at least 1909. By 1876, 20 miles of ditches had been completed, bringing water from the Elk River to the placers, and hydraulicking was to begin. In 1907 a very small dredge worked a few months on Willow Creek near its junction with Ways Gulch. In 1931-39 there were from 1 to 20 small operations each year. There have been sporadic small operations since.

Hahn's Peak is an igneous stock intruded into the sedimentary rocks which surround it. These sedimentary rocks include shales, sandstones and conglomerates. No important bedrock gold deposits have been found in the Hahns Peak district. The porphyry which makes up the stock and the sedimentary rocks immediately adjoining it carry varying, but small, amounts of gold and free gold has been found in porphyry from near the contact. The placer gravels contain predominantly sub-angular porphyry boulders from the stock, but also some rounded boulders of granite and gneiss. The porphryr came from Hahn's Peak but the rounded boulders of granite and gneiss were derived from one of the conglomerates.

Granite and gneiss occur at Farwell Mountain, six miles east of Hahns Peak village, and in the Park Range, six miles farther east. At both places quartz veins carrying a little gold have been found. Presumable these or similar sources provided the boulders found in the conglomerate. Most of the placer gold is fine--one millimeter or less--but some pieces of courser gold and some gold attached to quartz have been recovered. No gold-bearing quartz veins have been found in Hahns Peak.

With this limited, but conflicting, evidence it is not surprising that arguments have arisen about the source of the gold--whether Hahns Peak porphyry and marginal altered rocks, the older conglomerates or the veins in the mountains to the east are the immediate source. I believe that the porphyry and its marginal altered rocks or perhaps deposits related to them undiscovered or even now eroded away, are the immediate and the original sources of the placer gold. The deposits are in streams which radiate from the Hahns Peak intrusive. If the conglomerates were the source, since they are widespread there should be similar placers elsewhere and there are none. If the quartz veins of the Park Range were the immediate source, it is difficult to explain why there are no placers between Hahns Peak and that source.

Hahns Peak has not been actively glaciated, although it probably had hanging ice fields. There have been many landslides in its gulches and debris from them has accumulated at the foot of the peak. Placers have been traced and worked into these accumulations, but no placer deposits have been found above them.

The gold found in the placers is mostly small-scale than one millimeter in size. It is light yellow but contains considerable silver, fineness appears to be in the 600-750 range.

Select your samples at Poverty Bar or in Ways Gulch or on Willow Creek. Poverty Bar is a remnant of an alluvial fan. its gold will be concentrated in local channels at various levels; each channel should be marked by a lens of coarser fragments in the gravels. The other two are stream placers in which gold is concentrated near bedrock. On Willow Creek there may be skim bars. Reportedly, drilling on Willow Creek and lower Ways Gulch showed good values, but the bedrock gravels are probably too deep for you to reach. The placers in Little Red Park are colluvial deposits with chaotic gold distribution. They are difficult to find and to reach and aren't recommended for a first try.

SAN JUAN COUNTY

All the major valleys of San Juan County were glaciated in Wisconsin time and the placers they may have contained before were destroyed. The little placer gold they contain has been deposited in post-glacial, and even in modern, time. Although the county has been a major gold producer, placer mining has never been important here. Nevertheless the area can be of interest to the

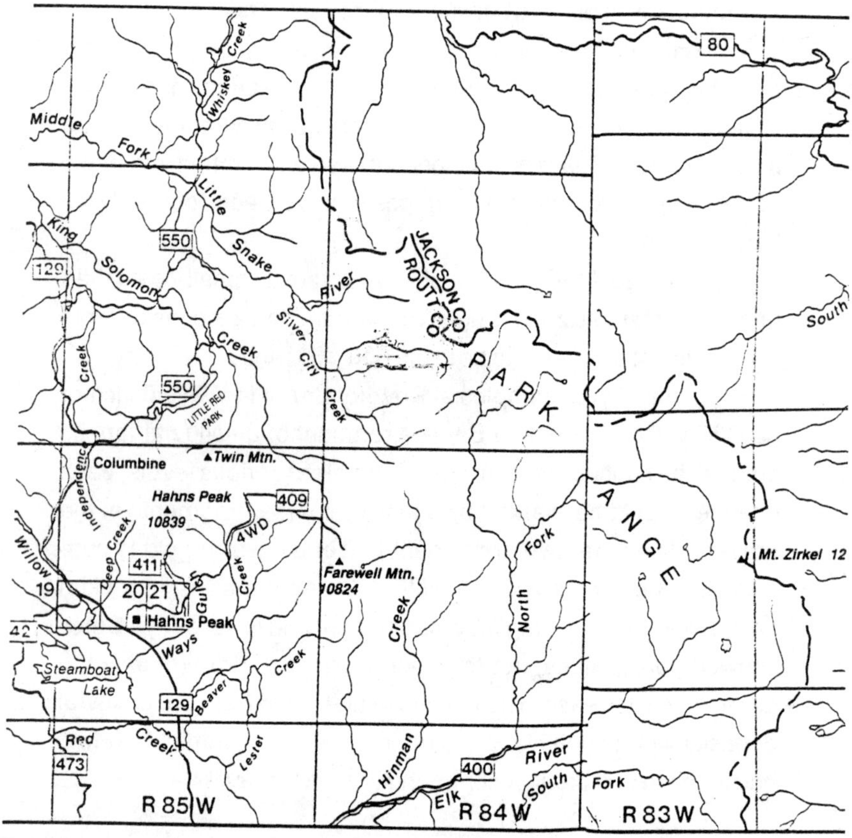

Figure 52. Index map of Hahns Peak and Independence Mountain placers, Routt and Jackson (

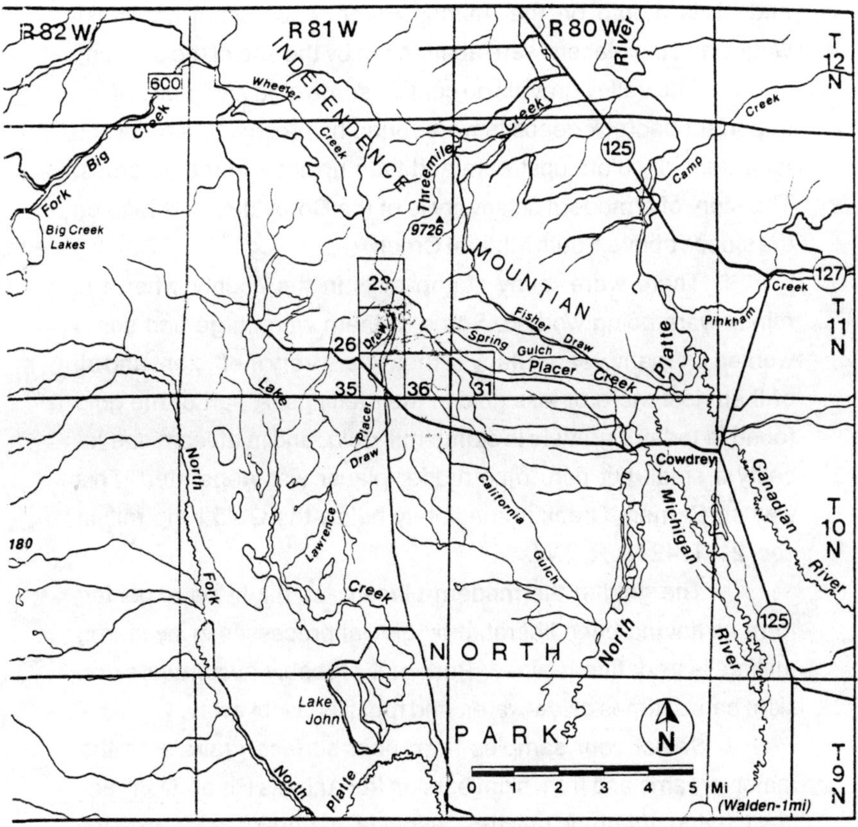

Counties. (Modified from U.S.G.S. topographic maps)

panner.

One of the first gold discoveries of the San Juan Mountains was made by Charles Baker's group of six men in 1860, near the site of the old town of Eureka in NE sec. 19, T. 42 N., R. 6 W., (unsurveyed). The following year a group of over 100 men arrived and placered in Eureka Gulch and at its mouth on the Animas River. They could not make wages and the placers were abandoned by the end of the season.

The valley gravels do contain some gold and many of the important bedrock deposits were found by prospectors following faint trails of colors upstream and then upslope to the outcrops. The story of a modern discovery is of the Corey deposits high on the slopes above South Mineral Creek.

There were many stamp mills in the county when the mines were being worked. Some of them were large and some worked for many years. Many of them were very inefficient and all of them lost more or less gold in their tailings. A part of the gold found in today's gravels is stamp mill gold, and in at least one locality a small but rich "man made" placer was generated. This was on Cement Creek immediately below the Gold King mill in sec. 21, T. 42 N., R. 7 W.

The gold in the modern placers, as might be expected from its having been liberated by glacial processes or by stamp milling, is very fine-grained. Park of it probably accumulates on skim bars at times of low water, and much is swept away.

Select your samples from near-surface gravels on the major streams and from skim bars on the Animas River. Don't neglect trying tributary gulches, especially Eureka and Arrastra Gulches. Below Silverton the gravels of the Animas River are gold-bearing and some of the lefthand tributaries (on the east side) downstream are worth prospecting. It's worthwhile to inquire about the old mills and check the slopes below them. The mills should have operated before World War I--most later mills lost less gold.

SAN MIGUEL AND MONTROSE COUNTIES

Individual placer operations years ago are reported on Big Bear and Bilk Creeks in T. 42 and 43 N., R. 10 W., and placer gold is said to have been found on the upper part of Mesa Creek drainage, probably in the southern part of T. 49 N., R. 17 W. The workings must be small for I was never able to locate them. The placer production of San Miguel- Dolores Counties has not been large, and almost all has come from the Keystone and San Miguel-Dolores River placers. Of these two, the Keystone placer has probably accounted for the greater part of the gold.

Some of the State's largest gold mines are located in the upper parts of the San Miguel and South Fork drainages. However, as is the case in Ouray and San Juan Counties, the gold placer deposits developed from them are not nearly so extensive as those derived from some much less productive districts. This is true of all the gold districts of the San Juan Mountains.

The first significant placer found going downstream from the head of the San Miguel River is the Keystone placer, at its junction with the South Fork, nearly four miles downstream from Telluride and more than five miles downstream from the nearest important bedrock gold deposits. Only a few small, very low-grade placers that did not pay to work were found by early prospectors on the flat east of town between the Keystone placer and Telluride. From place to place in the streams above Telluride and on the slopes they must have found colors leading to some of the vein outcrops, but they found no placer deposits worth working.

Probably there is now some modern stamp mill gold on skim bars, but significant modern placers should not be expected immediately below the old mills ass occurred at the Gold King mine in San Juan County. This is because the discharge streams of most of the mills here were so overloaded that placer concentration usually did not occur and also because when cyanide ex-

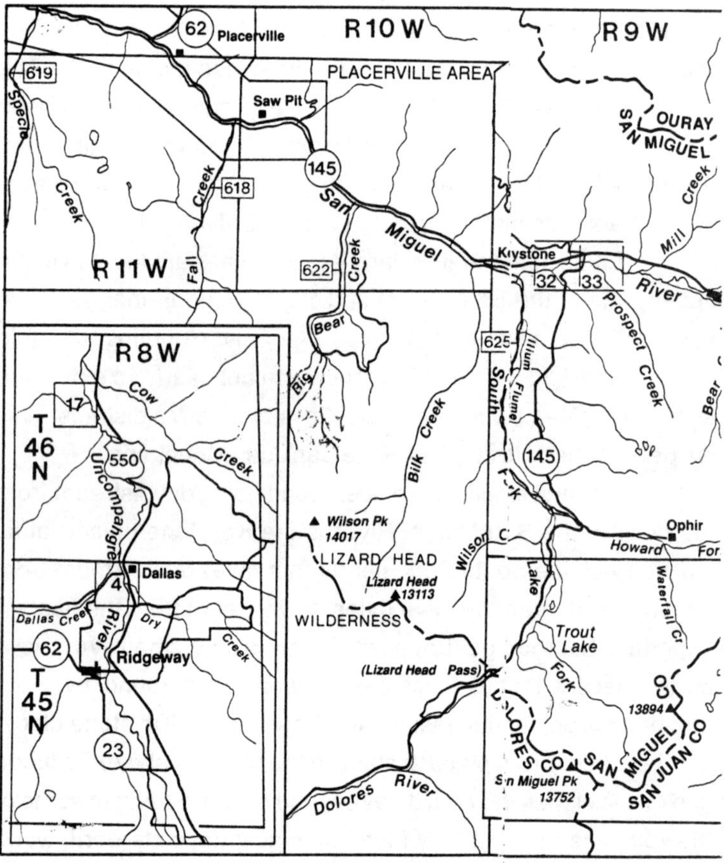

Figure 53. Index map of San Juan, Ouray, and San Miguel County placers. (Modified from U.S.

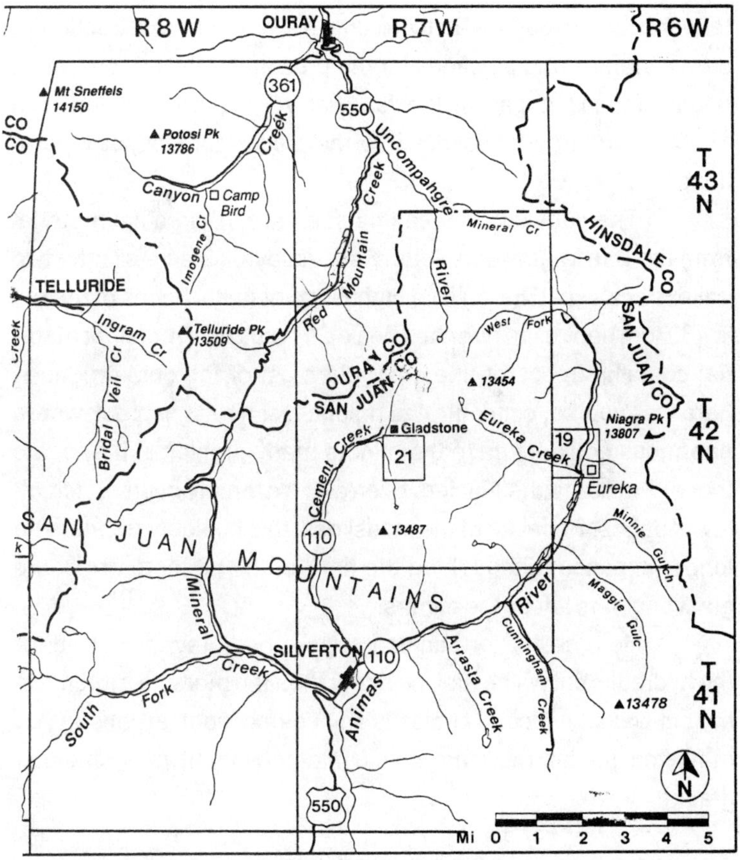

G.S. topographic maps)

traction of gold was developed, plants were built here that treated most of the accessible stamp mill tailings accumulations.

KEYSTONE PLACER

The Keystone placer is in the NE. sec. 32 and NW sec. 33, T. 43 N., R 9 W. The first work here was in 1881 and some drifting and small-scale hydraulicking continued each season until 1892. Afterward there was little activity until 1901 when construction of dams, glumes and pipelines for the present large cut's operation began. Mining began in the fall that year and continued until 1906, when large landslides into the placer cut stopped operations.

The cut walls that can be seen are in morainal material from the San Miguel and South Fork valleys. (Glaciers in the two valleys coalesced here.) The highest point on the faces of the cut is 430 feet above the river bed beside the cut. The morainal material contains traces of fine gold, but most of the gold produced here was coarse, concentrated on and just above bedrock where enormous boulders gave the miners many problems. It appears that at Keystone the glaciers overrode extremely course, rich alluvial debris. Because of the landslides, this bouldery debris is no longer exposed, although on the floor of the pit there are some giant boulders left by the miners.

Here select for panning stream bed gravels just below the hydraulic cut which may be richer than the moraine material in the cut walls. A second choice is any coarse, cobbley lenses you may find in the moraine which might be meltwater channel gravel.

SAN MIGUEL-DOLORES RIVER

The San Miguel River heads in Bridal Veil and Ingram Basins southeast of Telluride, flows through Telluride, Keystone, and Placerville in San Miguel County, flows through Naturita and Uravan in Montrose County and joins the Dolores River approximately three miles downstream from Utavan. In Montrose

County it is joined by Cottonwood Creek in sec. 3 T. 46 N., R. 14 W., and by Roc Creek in sec. 4, T. 48 M., R. 18 W.

Most of the placer work has been done in three areas; on the San Miguel downstream from Cottonwood Creek for about six miles and on the Dolores for nine miles upstream from Roc Creek. Placerville is near the center of the first group of placers; Naturita--several miles below the second group; and Uravan--just upstream from the third group.

The placers in the Placerville area probably were discovered in 1875. They were worked from 1876 through 1882 and occasionally thereafter. Probably this was because of the difficulty in getting water to the terraces or the gravels to water, for the gravels here are apparently higher grade than those of the placers downstream

In 1883 several companies were formed to work placers in the two downstream areas. One of these was the Montrose Placer Mining Co., with properties at Mesa Creek in the lowest area, which built the spectacular "hanging flume" in the Dolores River Canyon. (Remnants of the flume can still be seen in the canyon.) The company's water supply was provided by a ditch and flume over 10- miles long, about half ditch and half flume. It traversed the Dolores Canyon, a distance of about four miles, in a flume six feet wide and four feet deep pinned to the sheer rock of the canyon wall. In one stretch several hundred feet long where the wall overhangs, the flume hung from pins set in the rock overhead. The flume in the canyon was 100 to 150 feet above the river and several hundred feet below the edge of the canyon.

A road was built along the lip of the canyon and workmen hung from the top on ropes to mark the level of the flume on the wall. The holes were drilled and the flume constructed from a derrick on the flume floor that was advanced as flume sections were completed. A construction crew of 12 men was able to build up to 250 feet of flume per day. The ditch and flume were built upstream from the placer site, since it was nearest the lumber source.

The property was tested in the spring of 1888, the upper end of the ditch and flume was being constructed in 1890, construction of the tail sluice below the placer began early summer of 1891, the first cleanup was made about September 1891, and the Company suspended operations in 1893. At the time it was reported that the flume cost "something over $100,000," consumed 1.8 million board feet of lumber and was built in 'more than two years."

The Montrose Company was one of five companies that had projects at the Mesa and Cottonwood Creek areas and one of two that actually operated. It brought water to the Mesa Creek area; the U. S.. Gold Placers Company brought water to high bars opposite Cottonwood Creek in a nine-mile ditch. Both companies operated about a year and failed. In each cast the testing, if any, had been poorly done and the value of the placer gravels greatly exaggerated.

There was no further placering until the 1930's when there a number of small-scale operations on the rivers.

The river canyons are cut in sedimentary rocks with alternating hard and soft sections. In the three placer areas there are elevated terraces beside the river, each floored on hard sandstone beds at most places. There are many remnants of the terraces on the sloped of the canyons.

These terraces have been worked in cuts at many places, and in many of them the gravels can be seen or reached without too much cleaning. The terrace gravels have been worked in drift mines in many places. These drift mines are very old. They are unsupported, since they did not need timbering when they were driven, years ago. ***If you find any open, stay out of them. They are very dangerous!***

The gold in the river gravels is fine-grained. It is somewhat less so in the terrace gravels and less fine-grained at Placerville than downstream. The gold in the terrace gravels is concentrated near bedrock; toward the terrace surfaces the gravels are nearly barren. The gold in the stream gravels is concentrated in the upper few feet of gravel at skim bars. Bedrock beneath the modern river floors has never been tested.

Select gravel samples from skim bars along the modern rivers in the three areas described. You can also get samples for panning from as close to bedrock as possible in cuts on terraces. The table gives an idea how high to climb. There is no water on the terraces, so you must pack either gravel down or water up. I suggest the first. Once again, don't enter any drift workings!

SUMMIT COUNTY

BEMROSE PLACER AREA

Several placers have been worked in NE sec. 1 and central sec. 12, T. 8 S., R. 78 W., at the head of the Blue River near the foot of Hoosier Pass. In many places the gravels are covered by several feet of peat, which the miners had to strip. They are unusual in that the placers in sec. 12 are in moraine from the South Platte glacier that spilled over the top of Hoosier Pass. (On the south side of the pass the South Platte valley abruptly turns 90 degrees, flowing eastward from the high peaks of the Tenmile Range and then southward below the pass.) The placers in sec. 1 are in lateral moraine of the Monte Cristo Glacier that contains material from the South Platte Glacier.

MCNULTY GULCH

A small but apparently very rich placer, discovered in 1860, was worked in Mc Nulty Gulch in the S SE sec. 35, T. 7 S., R 79 W., and NE sec 2, T. 8 S., R. 79 W. from beneath the Climax tailings pond upstream almost three quarters of a mile. It is reported that bedrock gravels yielded as much as five ounces to the pan and that a 22 ounce nugget was found here in 1867, about at the edge of the pond. Gold occurs in thin gravels on bedrock and two to three inches into it. The gravel lies beneath glacial till in the lower

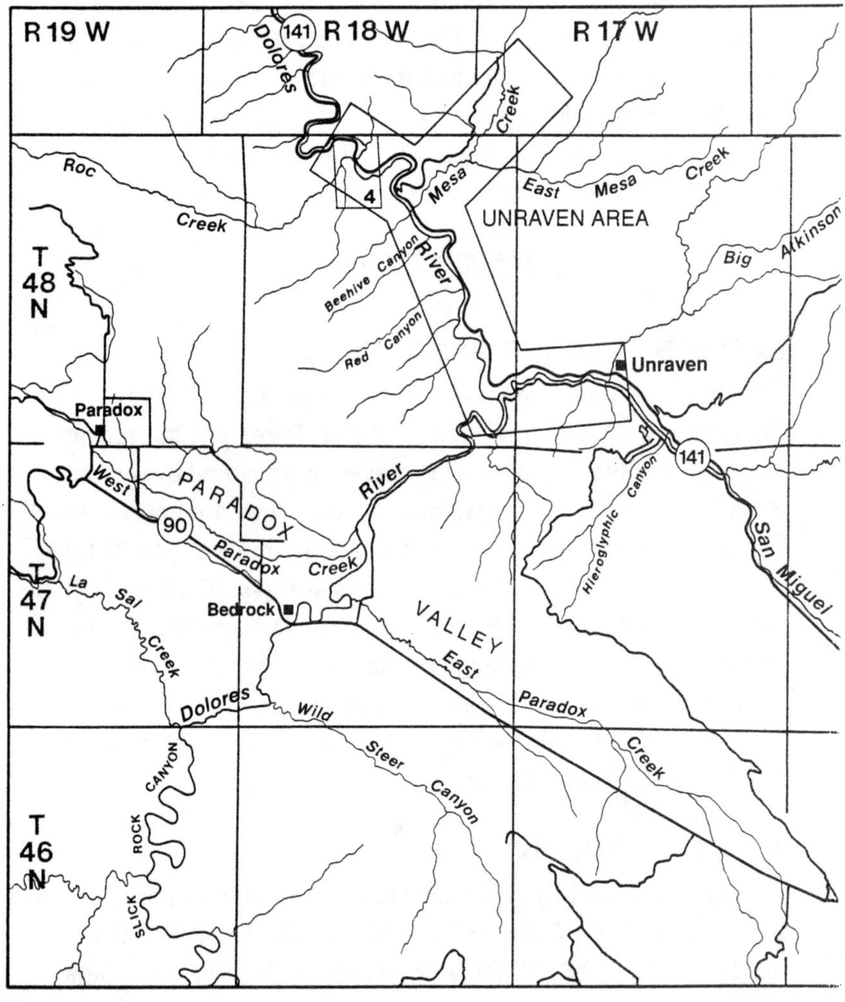

Figure 55. Index map of San Miguel-Dolores River placers Montrose County. (Modified from U.S.C

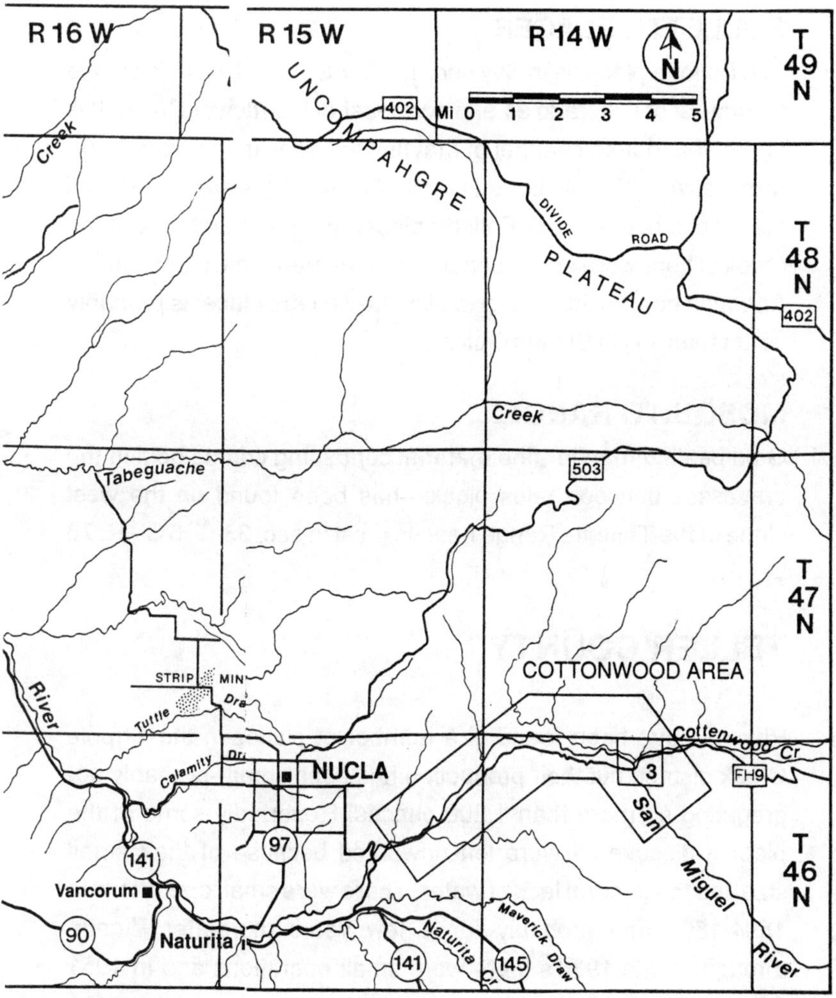

;.S. topographic maps)

part of the placer and was worked there by drift mining. The gold is coarse and not flattened. Its source has never been determined.

FOLLETTE PLACER

The Follette placer is in SW sec. 13, T. 7 S., R. 79 W. It lies to the southwest of Colorado 91 and northeast of Mayflower Creek. It is on an abandoned channel of Mayflower Creek in the glacial outwash terrace 75 feet above Tenmile Creek. There are some small placer pits between the Follette placer and the mouth of Clinton creek. There were some beneath the reservoir in Clinton Gulch. Although none is rich, the gravel of the Follette placer is probably better than any in Clinton Gulch.

MOSQUITO RANGE

Gold beating mantle--fine material containing coarse gold in the crevasses between talus blocks--has been found on the west slope of the Tenmile Range near its crest in sec. 33, T. 6 S., R. 78 W.

TELLER COUNTY

Placers have been found at a number of places in the Cripple Creek district, but their production has been small--probably aggregating no more than 1,200 ounces. Reportedly some of the placers discovered were left unworked because of their small size and the general lack of water. There were small operations in 1894-1897 and probably for a few years thereafter. Again, throughout the 1930's there were small operations and in 1951 there was a small dry land washing plant in operation.

The gold recovered was angular, mostly bright yellow but with some fragments with a rusty surface, and it contained very little silver.

The largest placer area was on the southwest slopes of

Mineral Hill and in Pony Gulch north and west of Cripple Creek town. Because the gravels are thin and the deposits spotty, the small placer dumps are not very noticeable. The gold clearly originated in Mineral Hill, but the bedrock sources have never been recognized.

Some placers were worked south of the town on Cripple Creek, and in Squaw and Arequa Gulches. The name "Gold Run" suggests that some placers were found there.

Placers were worked at Hull's Camp, near the center of SE sec. 20, T. 15 S., R. 69 W., and on Wilson Creek south of Victor There was an eluvial placer on Globe Hill in SE sec. 18, but its gravels were treated by milling.

Collect dirt for panning on the sloped of Mineral Hill and the head of Pony Gulch. This is the easiest area to locate and had the most extensive placers. From the foot of Mineral Hill up its slopes, the gold is in random distribution. In the broad valley of Pony Gulch and downstream it is concentrated on bedrock. Many of the other sites were once built up, and it will require some patience to find undisturbed gravel.

SELECTED REFERENCES

Conrow, Robert, 1983, The Great Diamond Hoax and Other True Tales, Boulder, Colo. Johnson Books, 62 p.

Ellis, E H., 1967, The Gold Dredging Boats Around Breckenridge, Colo., Boulder, Colo., Johnson Publishing Co 173 p.

Henderson, C. W., 1936. Mining in Colorado, A History of Discovery, development, and production: U.S. Geological Survey Prof. Paper 138, 263 p.

Hollister, O.J. 1867, The Mines of Colorado; Springfield, Mass. Samuel Bowles and Co., 450 p.

Parker, Ben H. Jr, 1961, The Geoology of the Gold Placers of Colorado, Colo. Sch. of Mines, Doctoral Thesis. 578 p.

______1974, Gold Placers of Colorado: Colorado School Mines Quarterly, v. 69, nos 3 and 4, p 268 and 224.

Peele, Robert, ed., 1941, Mining Engineers' Handbook, 3rd edition, v 1, John Wiley & Sons. 541 p.

Singewald, Q. D. 1950, Gold Placers and Their Geologic Environment in Northwestern Park Co. Colo. U.S. Geol Survey Bulletin 955-D, p. 103-172,

Wells, John H. 1969, reprinted with errata 1989, Placer Examination, Principles and Practice: U.S. Bureau of Land Management Tech. Bulletin. 4, 209 p.

Woodard, Bruce A., 1967, Diamonds In The Salt: Boulder Colo, Pruett Press. 200 p.

SUGGESTED READING

Chronic, J, and Chronic H. 1972, Prarie, Peak and Plateau--a guide to the Geology of Colorado; Colo Geol. Survey Bulletin 32, 126 p.

Davis, M. W. and Struefert, R. K., 1990, Gold Occurences of Colorado: Colo. Geological Survey Resource Series 28. 101 p.